PREFACE

This book was written for engineers and others who desire a good understanding of stack gas dispersion calculations. Anyone engaged in the design and environmental evaluation of air emission control systems will find the book to be a most valuable tool.

The technical literature on stack gas dispersion is quite extensive and dates back to the 1930's and earlier. Under the stimulus of stringent environmental control regulations, an immense growth in the use of dispersion calculation methods has occurred since the late 1960's. But prior to this book, there has been no single reference source that clearly and simply explains the fundamental theory and discusses the many assumptions and constraints involved in deriving the theory.[†] Unfortunately, most of the recent literature invariably references earlier publications and assumes that the reader has easy access to those earlier publications and fully understands them. All too often, those earlier publications also reference the reader even further back to much earlier publications which are extremely difficult to obtain. The result is that few readers actually trace the genesis of the fundamental theory.

This book was written to start "from scratch" and to derive the fundamental theory step-by-step, complete with many example calculations. All major facets of the theory are traced back to the early literature and developed in a simple, concise manner. As far as possible, esoteric advanced mathematics are avoided. However, the reader will need a good background in algebra and some knowledge of calculus.

Hopefully, this book will be useful as a single reference source that documents and explains the decades of development that led to today's highly sophisticated and computerized dispersion calculation models. Even more importantly, I hope that the reader will gain a clear understanding of the many idealized assumptions involved in the theory and methodology of dispersion calculations. For it is only with that understanding, that the reader can truly recognize and evaluate the inherent accuracy limits that are part and parcel of all our current dispersion calculation models.

As with any new book covering a complex subject, there may be some typographical errors, errors of omission, and perhaps errors of judgement. Any constructive suggestions for improvement of the book will be welcomed.

Irvine, California   1979, 1987, 1994   *Milton R. Beychok*

[†] With all due respect to Pasquill's classic text[1], which presumes an advanced knowledge in the theory of physics, kinetics and turbulence.

CONTENTS

Preface iii
List of Tables vii
List of Figures vii
List of Examples viii

Chapter 1: Atmospheric Parameters 1

The Atmosphere And Lapse Rates 1
Units Used In This Book 2
Pasquill Stability Classes 4
Potential Temperature Gradients 5
Inversion Layers And Mixing Heights 10

Chapter 2: Gaussian Dispersion Equations 19

Statistical Distributions 19
The Gaussian Distribution 21
Continuous Point-Source Plumes 24
Vertical Concentration Profiles 29
Crosswind Concentration Profiles 34
Discussion Of Input Parameters 38
Typical Plume Behavior Modes 40
Boundary-To-Centerline Concentration Ratios 44

Chapter 3: Dispersion Coefficients 45

The Historical Development Of Dispersion Coefficients 45
The Pasquill Dispersion Coefficients 46
Dispersion Coefficients In Equation Form 53
Rural Versus Urban Dispersion Coefficients 55

Chapter 4: Plume Rise 59

Introduction 59
Plume Rise Variables 60
Trajectory Of A Bent-Over Buoyant Plume 60
The Buoyancy Flux Parameter 61
The Stability Parameter 64
The Momentum Flux Parameter 67
Briggs' Equations For Bent-Over Buoyant Plumes 68
Calculated Plume Trajectories 73
Other Briggs Equations 81
Briggs' 1975 Lecture 82

Chapter 5:	**Time-Averaging Of Concentrations**	83
	The Components Of Turbulence	83
	Extrapolating Time-Averaged Concentrations	89
Chapter 6:	**Wind Velocity Profiles**	99
	The Need For Wind Velocity Profiles	99
	Wind Velocity Versus Altitude	100
Chapter 7:	**Calculating Stack Gas Plume Dispersion**	103
	Introduction	103
	Calculation Examples	104
	Example Graphs Of Ground-Level Concentrations Versus Downwind Receptor Distances	109
	Plotting Isopleths	112
	The Accuracy Of Dispersion Models	113
Chapter 8:	**Trapped Plumes**	119
	Introduction	119
	Multiple Reflections	120
	Where Vertical Dispersion Of Trapped Plumes Becomes Uniform	126
	Turner's Approximation	128
	Ground-Level Concentrations From Trapped Plumes	130
Chapter 9:	**Fumigation**	133
	Evolution Of A Plume Fumigation	133
	Ground-Level Concentrations During Fumigation	133
	Crosswind Dispersion During Fumigation	135
	Downwind Distance To Maximum Fumigation	136
	Calculation Example	141
	Time-Averaging Of A Fumigation	146
	Other Methods	147
	Closing Comment On Predicting Stack Gas Dispersion	148
Chapter 10:	**Meteorological Data**	151
	Wind Roses	151
	Distribution Of Calms	153
	The STAR Data	153
Chapter 11:	**Flare Stack Plume Rise**	167
	Introduction	167
	Flame Length	169
	Combusted Gas Temperatures	173
	Calculation Of Flare Stack Plume Rise	177

Chapter 12: Miscellany   179

    Short-Term Releases   179
    Source And Receptor At Different Ground-Level Elevations   183
    More Rigorous Definition Of Briggs' Buoyancy Parameter   183
    Converting Plume Concentration Expressions   185
    Effect Of Altitude On Ambient Air Standards   185
    Categorizing Atmospheric Stability With The
      Richardson Number   186
    More On Dispersion Coefficients In Equation Form   187

References   189

LIST OF TABLES

Table 1	Pasquill Stability Classes	8
Table 2	Mixing Heights Related To Stability Class	15
Table 3	Comparison Of Original Plume Spread Estimates	47
Table 4	Comparison Of Literature σ_z Values With Pasquill's Original Values	50
Table 5	Comparison Of Literature σ_y Values With Pasquill's Original Values	51
Table 6	Constants I, J and K For Use With Equation (27)	53
Table 7	Comparison Of Equation (27) With Figures 17 and 18	54
Table 8	Constants L, M and N For Use With Equation (28)	56
Table 9	Potential Temperature Gradients	67
Table 10	Conversion Of Time-Average Concentrations	92
Table 11	Conversion of Time-Average Concentrations	94
Table 12	Example Array of 24-hr Concentrations	97
Table 13	Exponents For Equation (52)	100
Table 14	Exponents For Equation (52)	101
Table 15	1-Hour Ground-Level Concentrations Calculated Under The Plume Centerline	117
Table 16	Downwind Distance At Which σ_z/L Becomes 1.2	128
Table 17	Solar And Sky Radiation Fluxes In The United States	139
Table 18	Recommended H_0 Values	140
Table 19	Optional Formats For The STAR Data	155
Table 20	Objective Definition Of Stability Classes	157
Table 21	A Typical Set Of STAR Data	159
Table 22	Calculated Adiabatic Flame Temperatures	175

LIST OF FIGURES

Figure 1	Atmospheric Conditions	6
Figure 2	Pasquill Stability Classes	7
Figure 3	Inversion Layers	11
Figure 4	Mixing Heights	14
Figure 5	Isopleths Of Mean Annual Morning Mixing Heights	16
Figure 6	Isopleths Of Mean Annual Afternoon Mixing Heights	17
Figure 7	Distribution Curve	20
Figure 8	Distribution Histogram	21
Figure 9	Continuous Point-Source Plume	25

Figure 10	Imaginary Plume Concept	25
Figure 11	Vertical Concentration Profile At Plume Centerline	31
Figure 12	Vertical Concentration Profile At Plume Centerline	32
Figure 13	Vertical Concentration Profile At Plume Centerline	33
Figure 14	Crosswind Concentration Profiles At Plume Centerline Heights	35
Figure 15	Crosswind Concentration Profiles From Plume Centerline To Ground Level	37
Figure 16	Plume Behavior	41
Figure 17	Vertical Dispersion Coefficients (Pasquill's rural)	48
Figure 18	Horizontal Dispersion Coefficients (Pasquill's rural)	49
Figure 19	Effect Of Urban Dispersion Coefficients	55
Figure 20	Vertical Dispersion Coefficients (comparing rural and urban)	57
Figure 21	Horizontal Dispersion Coefficients (comparing rural and urban)	58
Figure 22	Trajectory Of A Bent-Over, Hot Buoyant Plume	62
Figure 23	The Stability Parameter	65
Figure 24	Stack Gas Heat Emission vs Combustion Source Magnitude	74
Figure 25	Plume Trajectories Using Briggs' Equations	80
Figure 26	Smoothing A Continuous Wind Velocity Trace	87
Figure 27	Time-Averaged Plumes	88
Figure 28	Various Methods For Converting Time-Averaged Concentrations	93
Figure 29	Receptor Grid For Calculating 24-hr Average Concentration	96
Figure 30	Generalized Effect Of Altitude On Wind Velocity	102
Figure 31	Urban vs Rural σ Values	110
Figure 32	The Effect of Stack Height	111
Figure 33	Concentration Isopleths	114
Figure 34	Long-Term Ground-Level SO_2 Concentrations From A Stack Sited In Irregular Terrain	115
Figure 35	Sensitivity Study Of Gaussian Dispersion Models	118
Figure 36	The Concept Of Multiple Reflections	121
Figure 37	Vertical Dispersion Of Trapped Plumes Versus σ_z/L and H_e/L	127
Figure 38	Evolution Of A Plume Fumigation	134
Figure 39	Fumigation Parameters	145
Figure 40	A Typical Wind Rose	152
Figure 41	Plume Rise Parameters For Flare Stack Flames	168
Figure 42	Flame Length vs Heat Release	171
Figure 43	Combusted Gas Temperatures	176

LIST OF EXAMPLES

Example 1	Calculation Of Wet Adiabatic Lapse Rate	3
Example 2	Calculation Of Effective Stack Height	75
Example 3	Calculation Of Complete Plume Trajectories	76
Example 4	Calculation Of Complete Plume Trajectories	78
Example 5	Calculation Of Stack Gas Dispersion	105
Example 6	Calculation Of Stack Gas Dispersion	107
Example 7	Calculation Of Stack Gas Dispersion	108
Example 8	Fumigation Calculation	143
Example 9	Distribution Of Calms Using Data From Figure 40	154
Example 10	Calculation Of The Flame Height And The Buoyancy Factor Of The Combusted Gas Plume From A Flare Stack	178

Chapter 1

ATMOSPHERIC PARAMETERS

THE ATMOSPHERE AND LAPSE RATES

Atmospheric pressure decreases with increasing altitude, and any air rising from the warm surface of the earth will expand as it rises to lower atmospheric pressure levels. Taking the atmospheric pressure at sea level to be 14.696 psia (1,013 millibars[†]), we can obtain the atmospheric pressure at any altitude from this approximation:

(1) $P_a = 14.696(0.963)^a$

We can obtain the temperature which will be acquired by dry air rising from sea level to any given altitude from this expression which assumes that the rising air expands adiabatically:

(2) $T_a = T_s(P_a/14.696)^{(k-1)/k}$

where: a = altitude, in 1,000's of feet
 P_a = atmospheric pressure at a, in psia
 T_a = rising air temperature at a, in °R
 T_s = sea level ambient temperature, in °R
 k = 1.4 for dry air

Combining equations (1) and (2) for dry air:

(3) $T_a = T_s(0.963)^{0.286\,a}$

Assuming an ambient temperature T_s of 70 °F (530 °R), we can use the above equations to calculate:

Altitude (ft)	P_a (psia)	T_a (°R)	ΔP, psi (14.696 - P_a)	ΔT, °R (530 - T_a)	per 1000 ft ΔP	per 1000 ft ΔT
0	14.696	530	na	na	na	na
2000	13.629	519	1.067	11	0.53	5.5
4000	12.639	508	2.057	22	0.51	5.5
6000	11.721	497	2.975	33	0.50	5.5
8000	10.869	486	3.827	44	0.48	5.5

The above tabulation shows that, from sea level to an altitude of 8000 feet, the atmospheric pressure decreases about 0.5 psi per 1000 feet. It also shows that the temperature of rising dry air will decrease about 5.5 °F per 1000 feet.

[†] 1,000 millibars = 1 bar = 0.9869 atmospheres = 14.504 psia

Chapter 1: Atmospheric Parameters

The idealized adiabatic expansion and cooling of rising dry air by 5.5 °F per 1000 feet of rise is called the "dry adiabatic lapse rate". If the rising air contains any moisture (i.e., water vapor), the moisture will eventually condense as the rising air expands and cools. The resulting release of condensation heat will partially offset the adiabatic cooling of the air. Hence, the adiabatic lapse rate for moist air is less than for dry air.

Moist air rising from sea level (at 70 °F and 75 percent relative humidity) will cool to its dewpoint of about 60 °F when it reaches an altitude of about 1,800 feet. When it reaches 10,000 feet, about 45 percent of its original moisture will have condensed and the average "wet adiabatic lapse rate" will be about 3.4 °F per 1000 feet of rise (see Example 1).

The adiabatic lapse rate for air in motion (either rising or sinking) must not be confused with the actual ambient temperature gradient that may exist at a specific time in a specific location.[†] The actual ambient temperature gradient (i.e., the change in the ambient air temperature with increasing altitude) is a function of the time of day, the season of the year, the amount of solar radiation, the wind velocity, the rate of heat transfer from the ground surface to the ambient air, and many other factors.

Ambient temperature gradients are commonly obtained by using weather balloons equipped with temperature sensors and radio transmitters. The data obtained from weather balloons are called "temperature soundings" or "radiosondes". The measured ambient temperature gradient at any given time and location may be larger or smaller than the idealized adiabatic lapse rate at which rising air cools by expansion.

The basic reason for defining and deriving the dry and wet adiabatic lapse rates early in this chapter is to emphasize the fact that the lapse rates are based upon the idealized adiabatic cooling of rising air. As shown in subsequent sections of this chapter, comparison of the adiabatic lapse rates with the actual ambient temperature gradients provides a means for categorizing atmospheric turbulence.

UNITS USED IN THIS BOOK

Before reading Example 1, it should be made clear that this book purposely uses English and metric units. Using mixed units reflects the realities in the United States where typically: stack gas flow rates are available in pounds/hr or cubic feet/hr and not in grams/sec or cubic meters/sec; heat quantities are given as Btus and not as calories or joules; the Fahrenheit temperature scale is used rather than the Centigrade scale; pressures are measured in psi (pounds/square inch) and not as pascals or newtons/square meter; and windspeed data are measured in knots or miles/hr and not as meters/sec. On the other hand, most of the equations in this book use metric units. Thus, anyone involved in performing air dispersion calculations in the U.S. must learn to work with mixed units and become skilled in converting from English to metric units. Those who believe that all technical publications should use only S.I. metric units must accept the current realities in the United States.

[†] Much of the meteorological literature refers to the ambient air temperature gradient as an "ambient lapse rate" or sometimes as an "environmental lapse rate". Such designations are poor choices which can very easily lead to confusion with the "adiabatic lapse rate" for rising air.

Chapter 1: Atmospheric Parameters

EXAMPLE 1: CALCULATION OF WET ADIABATIC LAPSE RATE

GIVEN: 1000 cubic feet of air at sea level, 70 °F, 75 % relative humidity, specific heat (c_p) of 0.25 Btu/pound/°F.

SYMBOLS:
- P_a = atmospheric pressure at altitude a, psia
- P = water vapor pressure at the temperature of the air, psia
- pp = partial pressure of water vapor in the air, psia
- RH = percent relative humidity = (100)(pp/P)
- ACF = actual cubic feet of air and/or water vapor at the temperature and pressure of the air
- SCF = standard cubic feet of air and/or water vapor at 14.7 psia and 60 °F (520 °R)

(NOTE: 379 SCF is equivalent to 1 pound mol of air which is 29 pounds of air, or 1 pound mol of water vapor which is 18 pounds of water vapor)

AT SEA LEVEL:
  P = 0.363 psia at 70 °F (water vapor pressure taken from steam tables)
  pp = (RH/100)(P)   = (75/100)(0.363)   = 0.272 psia water vapor partial pressure
  water vapor in air = (0.272/14.7)(100) = 1.85 vol % = 18.5 ACF/(1000 ACF of wet air)

and thus:

°F	psia	ACF air	ACF water	SCF air	SCF water	pounds air	pounds water
70	14.7	981.5	18.5	963	18.15	73.7	0.862

AT 1,800 FEET ALTITUDE:
  $P_a = 14.696(0.963)^{1.8}$ = 13.73 psia atmospheric pressure
  $T_a = (70 + 460)(0.963)^{0.515}$ = 519.8 °R = 59.8 °F
  P = 0.254 psia at 59.8 °F (water vapor pressure from steam tables)

Assuming no condensation as yet, the air still contains 1.85 volume % of water vapor:
  pp = (1.85/100)(13.73)   = 0.254 psia water vapor partial pressure
  RH = (100)(0.254/0.254) = 100 percent

Thus, the air is at its dewpoint when at 59.8 °F, confirming the assumption of no condensation as yet. The air is now saturated and any more cooling will cause partial condensation of its water vapor content.

AT 10,000 FEET ALTITUDE:
  $P_a = 14.696(0.963)^{10}$ = 10.08 psia atmospheric pressure
  $T_a = (70 + 460)(0.963)^{2.86}$ = 475.8 °R = 15.8 °F

This T_a is based on the adiabatic expansion of dry air. However, the air is now below its dewpoint and some water has condensed out and released condensation heat which partially offsets the adiabatic cooling. Assume a 20 °F offset, giving a wet air temperature (T_{wa}) of 36 °F. Since the air is still saturated with water vapor even after partial condensation, RH is 100 percent and pp = P.

  T_{wa} = assumed 36 °F
  pp = P = 0.1041 psia at 36 °F (from steam tables)
  water vapor remaining in the air
    = (0.1041/10.08)(100)   = 1.033 volume %
  SCF of water vapor remaining in the air
    = [963/(1-0.01033)]-963   = 10.05 SCF
  SCF of condensed water vapor
    = 18.15-10.05   = 8.10 SCF
  pounds of condensed water
    = 8.10(18/379)   = 0.385 pounds
  condensation heat release at 1000 Btu/pound   = 385 Btu
  the resulting temperature offset
    = 385/[73.7(0.25)]   = 20.9 °F

This confirms the assumed 20 °F offset well enough, and thus:
  % of original water remaining in the air at 10,000 feet altitude
    = (10.05/18.15)   = 55 %
  average wet adiabatic lapse rate during the 10,000 feet rise
    = (70-36)/10   = 3.4 °F/1000 feet

PASQUILL STABILITY CLASSES

The amount of turbulence in the ambient air has a major effect upon the rise of stack gas plumes and upon the subsequent dispersion of the plumes. The amount of turbulence can be categorized into defined increments or "stability classes". The most widely used categories are the Pasquill Stability Classes A, B, C, D, E and F. Class A denotes the most unstable or most turbulent conditions, and Class F denotes the most stable or least turbulent conditions.

Atmospheric air turbulence is created by many factors, such as: wind flow over rough terrain; trees or buildings; migrating high and low pressure air masses and "fronts" which cause winds; thermal turbulence from rising warm air; and many others.

Any factor which enhances the vertical motion of air (either rising or sinking) will increase the degree of turbulence. The difference between the dry adiabatic lapse rate and a given ambient air temperature gradient provides a direct indication of whether vertical air motion will be enhanced or dampened. Thus, as discussed below, comparison of adiabatic lapse rates with ambient air temperature gradients can be used to define stability classes which categorize and quantify turbulence.

Super-adiabatic: Comparing an ambient air temperature gradient of -10 °F/1000 feet with the dry adiabatic lapse rate of -5.5 °F/1000 feet (see Figure 1):

Any rising air parcel (expanding adiabatically) will cool more slowly than the surrounding ambient air. At any given altitude, the rising air parcel will still be warmer than the surrounding ambient air and will continue to rise. Likewise, descending air (compressing adiabatically) will heat more slowly than the surrounding ambient air and will continue to sink because, at any given altitude, it will be colder than the surrounding ambient air. Therefore, any negative ambient air temperature gradients with larger absolute values than 5.5 °F/1000 feet will enhance turbulent motion and result in unstable air conditions. Such ambient air gradients are called "super-adiabatic" (more than adiabatic).

Sub-adiabatic: Comparing an ambient air temperature gradient of -4 °F/1000 feet with the dry adiabatic lapse rate of -5.5 °F/1000 feet, we might conclude that the ambient air gradient is less than the adiabatic lapse rate. However, as discussed earlier herein, humid air has a wet adiabatic lapse rate that is less than the dry adiabatic lapse rate. In fact, Example 1 illustrated a wet adiabatic lapse rate of -3.4 °F/1000 feet. Thus, a sub-adiabatic condition (illustrated in Figure 1) requires an ambient air temperature gradient of about -3 °F/1000 feet to be fairly certain that the ambient air gradient is less than the wet adiabatic lapse rate:

Any air parcel in vertical motion (expanding or compressing adiabatically) will change temperature more rapidly than the surrounding ambient air. At any given altitude, a rising air parcel will cool faster than the surrounding air and tend to reverse its motion by sinking. Likewise, a sinking air parcel will warm faster than the surrounding air and tend to reverse its motion by rising. Thus, negative ambient air temperature gradients with lower absolute values than 3 °F/1000 feet will suppress turbulence and promote stable air conditions. Such ambient air gradients are called "sub-adiabatic" (less than adiabatic).

Inversions: A positive ambient air temperature gradient is referred to as an "inversion" since the ambient air temperature increases with altitude (see Figure 1). The difference between

Chapter 1: Atmospheric Parameters 5

the positive ambient air gradient and either the wet or dry adiabatic lapse rate is so large that vertical motion is almost completely suppressed. Hence, air conditions within an inversion are very stable.

Neutral: If the ambient air temperature gradient is essentially the same as the adiabatic lapse rate, then rising or sinking air parcels will cool or heat at the same rate as the surrounding ambient air. Thus, vertical air motion will neither be enhanced nor suppressed. Such ambient air gradients are called "neutral" (neither more or less than adiabatic).

Figure 1 illustrates three types of atmospheric stability conditions, namely: super-adiabatic, sub-adiabatic and inversion. The entire spectrum of atmospheric conditions (ranging from super-adiabatic to inversion) is divided into the six Pasquill stability classes in Figure 2, with each class defined as being within a specific range of ambient air temperature gradients. Table 1 includes those temperature gradient ranges as well as other parameters commonly used in defining the Pasquill stability classes.

This section has focused upon the Pasquill stability classes for categorizing the intensity of atmospheric turbulence. However, there are many other methods which have been used for categorizing atmospheric turbulence. Some of the other methods make use of:

- The standard deviation of angular wind fluctuations (see Chapter 5).
- The vertical heat flux (H_o) derived from heating of the ground surface by solar radiation (see Chapter 9).
- The Richardson number (R_i) as discussed in Chapter 12.
- The potential temperature gradient ($d\theta/dz$) as discussed in the next section of this chapter.

The Pasquill stability classes will be used throughout the subsequent sections of this book because they have gained wide usage and acceptance and because they are relatively easy to determine.

POTENTIAL TEMPERATURE GRADIENTS

Parameters other than those presented in Table 1 are sometimes used to define atmospheric stability classes. For example, the TVA (i.e., the Tennessee Valley Authority) prefers to use the "potential temperature gradient ($d\theta/dz$)" for defining atmospheric stability classes.[2] Basically, the potential temperature gradient is the difference between the ambient air temperature gradient and the dry adiabatic lapse rate:

(4) $d\theta/dz$ = dT/dz - dT_a/dz
              = dT/dz + 5.5 °F/(1000 ft)

$$\begin{aligned}
\text{where:} \quad d\theta/dz &= \text{potential temperature gradient, °F/(1000 ft)} \\
dT/dz &= \text{ambient air temperature gradient, °F/(1000 ft)} \\
dT_a/dz &= \text{dry adiabatic lapse rate, °F/(1000 ft)} \\
&= -5.5 \text{ °F/(1000 ft)} \\
&= -10 \text{ °C/km} \\
z &= \text{altitude} \\
\text{gradient} &= \text{rate of temperature change with increasing altitude}
\end{aligned}$$

6 Chapter 1: Atmospheric Parameters

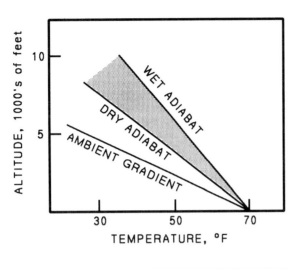

SUPER-ADIABATIC CONDITION

Ambient temperature gradient is negative and absolute value is greater than 5.5 °F/1000 feet.

Turbulence is enhanced and the air is unstable.

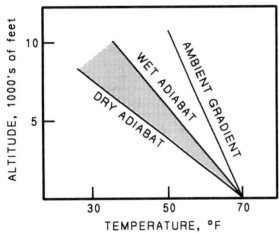

SUB-ADIABATIC CONDITION

Ambient temperature gradient is negative and absolute value is less than 3.0 °F/1000 feet.

Turbulence is suppressed and the air tends toward being stable.

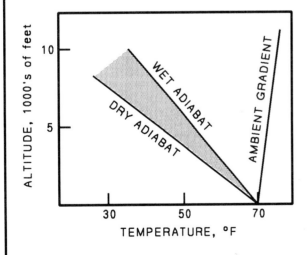

INVERSION CONDITION

Ambient temperature gradient is positive.

Turbulence is almost completely suppressed and the air is very stable.

FIGURE 1

ATMOSPHERIC CONDITIONS

FIGURE 2
PASQUILL STABILITY CLASSES

PASQUILL STABILITY CLASS	AMBIENT TEMPERATURE GRADIENT [3] (°F/1000 feet)
A -- very unstable	less than -10.4
B -- unstable	-10.4 to -9.3
C -- slightly unstable	-9.3 to -8.2
D -- neutral	-8.2 to -2.7
E -- slightly stable	-2.7 to +8.2
F -- stable	more than +8.2

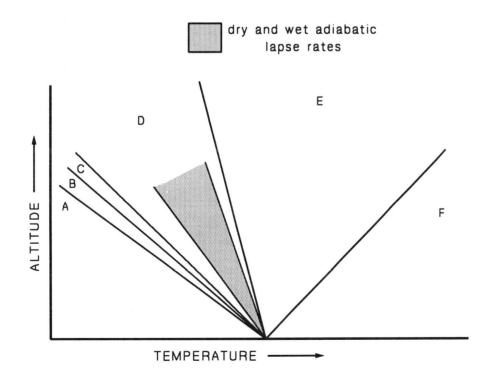

TABLE 1

PASQUILL STABILITY CLASSES

PASQUILL STABILITY CLASS	AMBIENT TEMPERATURE GRADIENT[3] (°F/1000 feet)
A -- very unstable	less than -10.4
B -- unstable	-10.4 to -9.3
C -- slightly unstable	-9.3 to -8.2
D -- neutral	-8.2 to -2.7
E -- slightly stable	-2.7 to 8.2
F -- stable	more than 8.2

| PASQUILL STABILITY CLASS RELATED TO WINDSPEED AND INSOLATION[a] |||||||
| Surface Wind || Day-time Insolation[a] ||| Night-time Cloud Cover[b] ||
m/sec	mi/hr	strong	moderate	slight	> 4/8 cloud	< 3/8 cloud
<2	<5	A	A-B	B	---	---
2-3	5-7	A-B	B	C	E	F
3-5	7-11	B	B-C	C	D	E
5-6	11-13	C	C-D	D	D	D
>6	>13	C	D	D	D	D

[a] insolation[3] (incoming solar radiation):
  Strong > 143 cal/m²/sec (or upward heat flux > 38 cal/m²/sec)
  Moderate = 72-143 cal/m²/sec
  Slight < 72 cal/m²/sec (or upward heat flux < 18 cal/m²/sec)

[b] Neutral class D applies to heavily overcast skies, day or night.

If a parcel of dry air at a given altitude above sea level were to be adiabatically compressed to a pressure of 1013 millibars (essentially sea level atmospheric pressure), the air parcel's temperature would then be:

(5) $\quad \theta = T_{sea\ level} + (dT/dz)\Delta z + (5.5)\Delta z$

Thus, assuming a sea level ambient air temperature of 70 °F and an ambient air temperature gradient of -10 °F/(1000 ft), a parcel of air from 2000 feet altitude above sea level moving downward to sea level could <u>potentially</u> acquire a temperature of:

$$\theta = 70 + (-10/1000)(2000) + (5.5/1000)(2000)$$
$$= 61\ °F$$

and:

$$d\theta/dz = -10 - (-5.5)$$
$$= -4.5\ °F/(1000\ ft)$$

As discussed earlier in this chapter and as shown in Figure 1, the difference between the ambient air temperature gradient and the dry adiabatic lapse rate can be used to categorize the degree of vertical motion turbulence in the atmosphere. Thus, the potential temperature gradient $d\theta/dz$ can be used to define stability classes just as conveniently as can be done by using the ambient air temperature gradient dT/dz.

The TVA uses potential temperature gradients to define their atmospheric stability classes, which are <u>not</u> the same as the more commonly used Pasquill stability classes.

The TVA's definition of their six stability classes using the potential temperature gradient $d\theta/dz$ are listed here along with the comparable equivalent values of the ambient temperature gradient dT/dz:

TVA Stability Class[2]	$d\theta/dz$ (°F/1000 ft)	Equivalent dT/dz[†] (°F/1000 ft)
Neutral	0.0	-5.5
Slightly stable	1.5	-4.0
Stable	3.5	-2.0
Isothermal	5.5	0.0
Moderate inversion	7.5	2.0
Strong inversion	9.5	4.0

[†] Obtained by using $dT/dz = d\theta/dz - 5.5$ (all in units of °F/1000 ft)

The TVA stability classes are compared below with the Pasquill stability classes as defined in Table 1 and Figure 2:

Pasquill Class	Average Ambient Gradient, °F/1000 ft†		TVA Class[2]
A, very unstable	< -10.4		
B, unstable	-9.9		
C, slightly unstable	-8.8		
D, neutral	-5.5	-5.5	neutral
		-4.0	slightly stable
		-2.0	stable
		0.0	isothermal
		2.0	moderate inversion
E, slightly stable	2.8		
		4.0	strong inversion
F, stable	> 8.2		

As another point of comparison, many of the U.S. EPA's dispersion models use these values for stable atmospheric conditions:

	dT/dz (°F/1000 ft)	dθ/dz (°F/1000 ft)
Pasquill Class E	5.5	11.0
Pasquill Class F	13.7	19.2

The concept of the potential temperature gradient $d\theta/dz$ is an important one, particularly in defining the behavior of buoyant stack gas plumes (see Chapter 4). However, to avoid any confusion, all further discussion of stability classes will be based on the Pasquill stability classes as defined by the ambient air temperature gradients dT/dz presented in Table 1.

INVERSION LAYERS AND MIXING HEIGHTS

The ambient air temperature gradient at any given time or location may change with altitude. In fact, it may change from positive to negative or from negative to positive. In other words, the atmosphere may have "inversion layers" within which the ambient temperature increases with altitude. As illustrated in Figure 3, a "surface inversion" may coexist with either a super or sub-adiabatic atmosphere above the surface inversion layer. Similarly, as also shown in Figure 3, an "inversion aloft" may coexist with atmospheres (above and below the inversion layer) which are either super or sub-adiabatic.

As previously noted, the atmosphere within an inversion is very stable with very little vertical motion. Any rising air parcel within the inversion layer soon expands and cools to a temperature lower than the surrounding ambient air (which increases with altitude within an inversion) and the air parcel stops rising. Any sinking air parcel within the inversion soon

---

† Multiply by 1.823 to obtain as °C/km

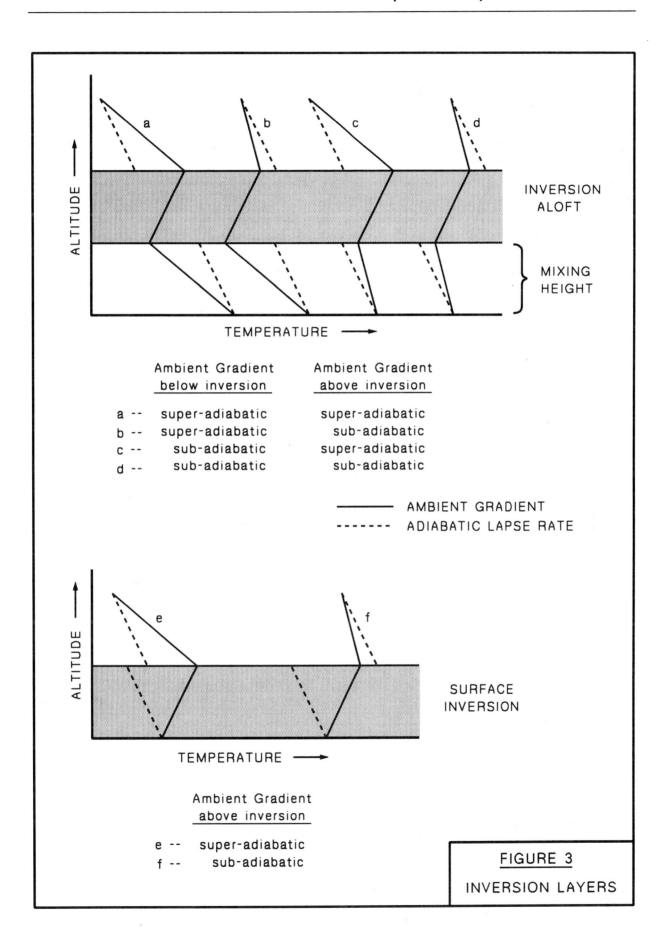

FIGURE 3
INVERSION LAYERS

compresses and heats to a temperature higher than the surrounding ambient air (which decreases with lower altitude) and stops sinking. For the same reasons, any stack gas which enters an inversion layer tends to be contained within the inversion and will undergo very little vertical mixing unless it has sufficient momentum to penetrate and pass through the inversion layer.

If the atmosphere above an inversion is super-adiabatic, any sinking air parcel reaching the inversion layer will be cooler than the top of the inversion layer (see a, c and e in Figure 3 and follow the adiabatic lapse lines down to the inversion) and will continue to sink. Thus, sinking air from a super-adiabatic atmosphere above an inversion tends to penetrate the inversion. Sinking air from a sub-adiabatic atmosphere above an inversion (see b, d and f in Figure 3) tends to not penetrate the inversion.

If the atmosphere below an inversion is super-adiabatic, any rising air parcel reaching the inversion layer will be hotter than the bottom of the inversion layer (see a and b in Figure 3 and follow the adiabatic lapse lines up to the inversion) and will continue to rise. Thus, rising air from a super-adiabatic atmosphere below an inversion tends to penetrate the inversion. Rising air from a sub-adiabatic atmosphere below an inversion (see c and d in Figure 3) tends to not penetrate the inversion.

From the foregoing discussion of inversion layers to this point, we can summarize that:

- Any stack gas entering an inversion layer tends to be kept within the inversion unless it has sufficient momentum to pass through the inversion.

- Air parcels descending from a super-adiabatic atmosphere above an inversion or rising from a super-adiabatic atmosphere below an inversion tend to penetrate the inversion.

- Air parcels from a sub-adiabatic atmosphere above or below an inversion tend not to penetrate the inversion.

Surface inversion layers may exist during the day or the night and they will usually result from either of these conditions:

- During the night, the earth's surface loses heat by radiation and the ground cools rather rapidly. The air above the surface cools less rapidly by convection (which is the meteorological term for heat transfer occurring from vertical air motion) so that the ambient air temperature some distance above the ground is higher than the air temperature near the ground. Thus, a surface inversion is formed in which the ambient air temperature increases with increasing altitude. This so-called "radiation inversion" extends up to some altitude at which the atmosphere reverts to its normal negative ambient gradient (i.e., the temperature decreases with increasing altitude).

- The lower air may have crossed a cold surface, such as a lake, and been cooled by advection (which is the meteorological term for heat transfer occurring from horizontal air motion) so that the lower air is cooler than the upper air. This forms a surface inversion known as an "advective inversion". A surface inversion occurring during the afternoon is apt to be an advective inversion.

During a typical diurnal pattern (i.e., daily cycle), the base of a radiation inversion formed during the night rises during the day as the earth's surface warms up. As the inversion rises, it forms an inversion aloft. The base of the inversion aloft forms a ceiling, or "lid", above which very little or essentially no vertical turbulence or vertical mixing occurs within the inversion layer.

Any stack gas plume dispersing in the atmospheric layer beneath the inversion base will be limited in vertical mixing to that which occurs beneath the inversion "lid". Even if the stack gas plume penetrates the inversion, it will not undergo any significant additional vertical mixing with the air within the stable inversion layer. Thus, the available vertical "mixing height" is defined as the distance from the earth's surface to the base of any inversion aloft (see Figure 3).

As the day goes on and the earth's surface continues to warm up, the base of the inversion layer rises, the vertical mixing height becomes higher, and the inversion layer becomes thinner. When the inversion base reaches the inversion top (perhaps in the midafternoon on a hot summer day), the inversion layer breaks up completely and the mixing height then becomes unlimited.

The inversion layer may also be thinned from the top downward by vertical turbulence and penetration of colder air from above the layer. However, that would not increase the mixing height below the inversion layer.

It should be noted that, although the mixing height is limited by an inversion aloft, surface winds and turbulence beneath the inversion layer will still provide vertical and crosswind mixing. Even within an inversion layer, some crosswind mixing is still available for dispersion of any stack gas plume which enters the inversion layer.

As a final note on inversion layers, inversions aloft may be formed by conditions other than the day-time heating of the base of a radiation inversion formed during the previous night. Upper air frontal disturbances and high pressure areas can form inversions aloft known as "frontal" or "subsidence" inversions.

Ideally, the available mixing height beneath an inversion aloft at any time is best determined from an ambient air temperature gradient (see upper part of Figure 3) obtained at the specific time and location. However, a single ambient air temperature gradient obtained during the day can be used to estimate both the morning and afternoon mixing heights. Assuming that any change in the ambient air temperature gradient is caused solely by heating at the earth's surface, and further assuming that the atmospheric layer beneath the inversion aloft has an ambient temperature gradient approaching the dry adiabatic lapse rate:

- The morning mixing height is defined by the height at which a dry adiabatic lapse rate line drawn from the estimated morning temperature at the earth's surface intersects the ambient air temperature gradient line.

- The afternoon mixing height is defined by the height at which a dry adiabatic lapse rate line drawn from the estimated afternoon temperature at the earth's surface intersects the ambient air temperature gradient line.

Figure 4 depicts the above concept and illustrates how the dry adiabatic lapse rate can be utilized along with an ambient air temperature gradient obtained at 0600 hours to determine the available mixing heights at 0400, 0600, 0900 and 1500 hours:

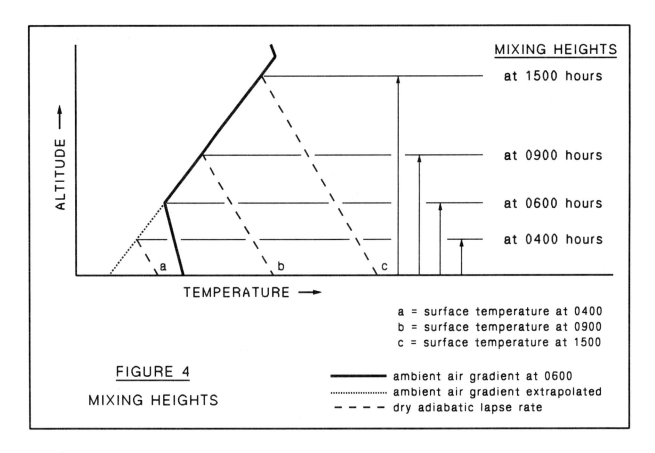

FIGURE 4

MIXING HEIGHTS

a = surface temperature at 0400
b = surface temperature at 0900
c = surface temperature at 1500

——— ambient air gradient at 0600
.... ambient air gradient extrapolated
— — — dry adiabatic lapse rate

Holzworth[4] used the above concept to determine mean annual and seasonal morning and afternoon mixing heights with data collected by the National Climatic Center and others from 62 weather stations for the years 1960 through 1964. Holzworth then developed 10 maps of the contiguous United States with isopleths of the:

- Mean annual morning mixing heights
- Mean winter morning mixing heights
- Mean spring morning mixing heights
- Mean summer morning mixing heights
- Mean autumn morning mixing heights
- Mean annual afternoon mixing heights
- Mean winter afternoon mixing heights
- Mean spring afternoon mixing heights
- Mean summer afternoon mixing heights
- Mean autumn afternoon mixing heights

Holzworth's isopleth maps for the mean annual morning and afternoon mixing heights are presented herein as Figures 5 and 6.

Holzworth's mixing heights are based on ambient temperature gradients that were obtained at 1200 GMT which corresponds to 0400 PST, 0500 MST, 0600 CST and 0700 EST. His morning mixing heights used surface temperatures measured at hourly intervals from 0200 through 0600 Local Standard Time (LST) and included a plus 5 °C to allow for the difference between urban surface temperatures and the usual rural or suburban locations of the weather stations. Therefore, Holzworth states "the <u>urban</u> morning mixing height was calculated".

Holzworth also tabulates the mean annual and seasonal mixing heights for each of the 62 locations in his Table B-1.[4] This is an example excerpted from that table:

		Annual Mean				
		H			U	
Station	Time	NOP	All	% NOP	NOP	All
Albany, NY	M	538	675	72.1	4.5	5.4
	A	1414	1484	72.3	7.2	7.6

where: M = morning
A = afternoon
H = mixing height, m
U = windspeed in the mixing layer, m/sec
NOP = during non-precipitation periods
All = includes precipitation and cold air periods, and missing data points

In those cases where stack gas dispersion is limited by the mixing height, it is useful to relate mixing heights to a shorter term parameter. One U.S. EPA dispersion model[5] does just that by correlating Holzworth's mixing heights with the Pasquill stability classes as follows:

TABLE 2

MIXING HEIGHTS RELATED TO STABILITY CLASS

Stability Class	Mixing Height, meters
A	1.5 × AMH
B	AMH
C	AMH
D (day)	AMH
D (night)	½ (AMH + MMH)
E	MMH
F	MMH

(AMH and MMH are the afternoon and morning mixing heights)

The EPA's correlation appears to be a logical and reasonable approach which recognizes that:

- Very unstable surface conditions (stability class A) should tend to erode any inversion aloft and, therefore, to raise the mixing layer lid.

- Very stable surface conditions (stability classes E and F) occur mostly at night, and Holzworth's morning mixing height (0200 to 0600 LST) is representative of the nocturnal mixing height.

Note: Data on mixing heights in Canada have been published by Portelli.[6]

16   **Chapter 1: Atmospheric Parameters**

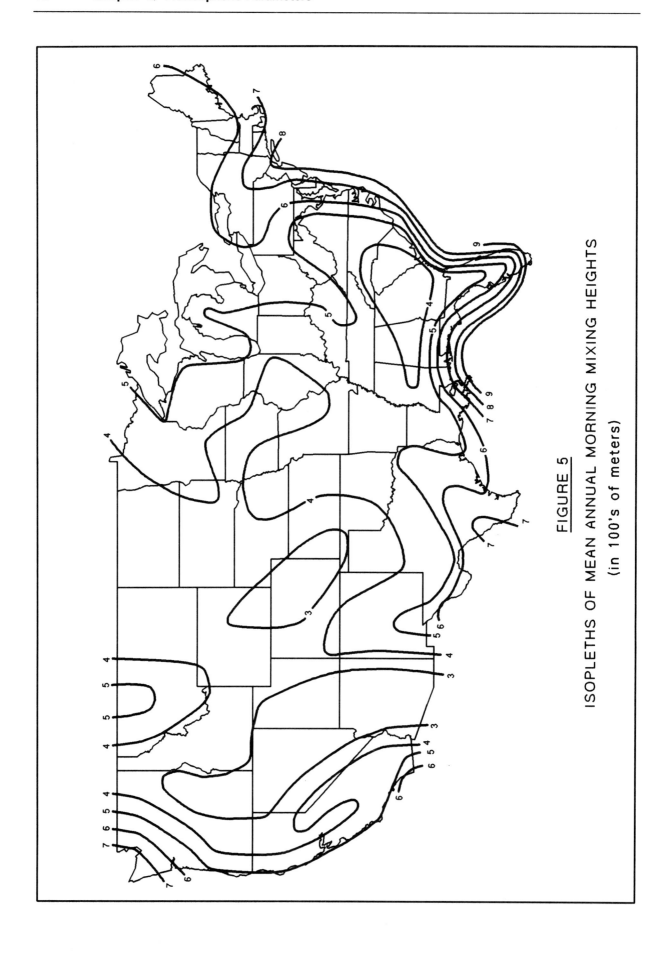

FIGURE 5
ISOPLETHS OF MEAN ANNUAL MORNING MIXING HEIGHTS
(in 100's of meters)

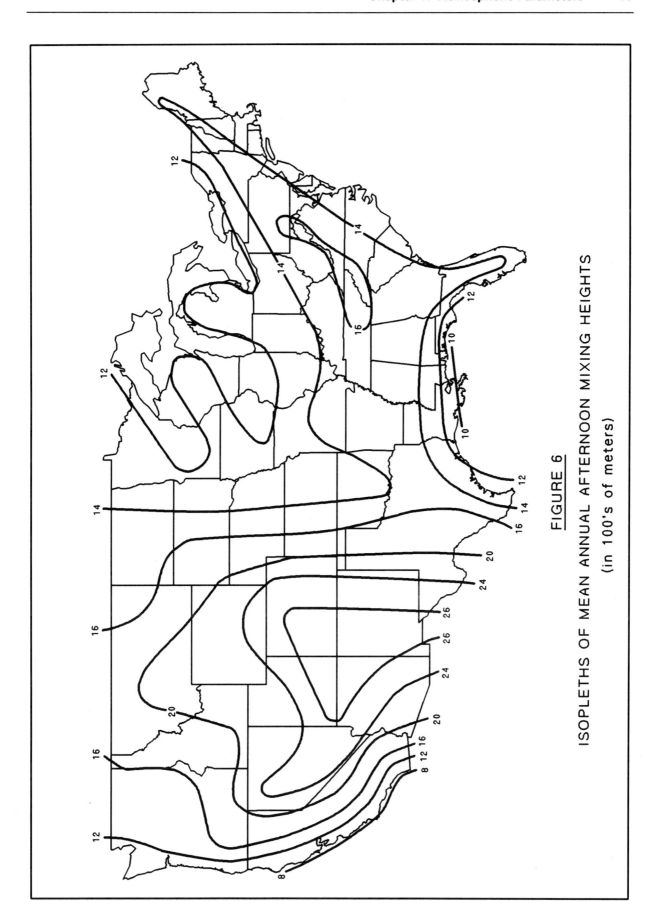

FIGURE 6
ISOPLETHS OF MEAN ANNUAL AFTERNOON MIXING HEIGHTS
(in 100's of meters)

Chapter 2

GAUSSIAN DISPERSION EQUATIONS

STATISTICAL DISTRIBUTIONS

Let us consider the population of a large class of objects, with each class member possessing some amount of a measurable characteristic. Then let us divide the population into smaller subclasses wherein each member of any specific subclass has the <u>same amount</u> of the measurable characteristic (or within a small increment of the same amount). Now examine how the total population would be distributed among the various subclasses.

The following definitions and examples should hopefully clarify the terminology involved:

- N = total population or members of a large class (dimensions: members)
- x = a measurable characteristic possessed in some amount by all of the members of the large class (dimensions: d)
- x_i = a specific amount of the measurable characteristic (dimensions: d)
- λ = a defined increment or interval of the measurable characteristic (dimensions: d/interval)
- n = the population of any subclass contained within the defined interval of the measurable class characteristic (dimensions: members/interval)
- n_i = the population of the specific subclass which has x_i amount of the measurable class characteristic at the midpoint of its defined interval of measurable class characteristic (dimensions: members/interval)

Some examples using the above definitions are:

N, class population	x, measurable characteristic	λ	x_i	n_i
N soldiers	chest sizes	1"	43"	n_i soldiers in a chest size range of 42.5 to 43.5 inches
N students	test scores	4 units	80 units	n_i students in a score range of 78 to 82 units
N homes	$ value	$2,000	$30,000	n_i homes in a value range of $29,000 to $31,000
N wastewaters	oil content	2 ppm	50 ppm	n_i wastewaters in an oil content range of 49 to 51 ppm
N gas molecules	distance from release center	1 m	200 m	n_i molecules at a distance range from the release center of 199.5 to 200.5 meters

When dealing with a large class population grouped into smaller subclasses, it is important that n_i values be referenced to a defined λ interval, because n_i is the population of a <u>defined size</u> of a subclass <u>located at a specific place</u> on the scale of class characteristic. In other words:

- n_i is the population of a specific subclass.

- λ is the size of the subclass on the linear scale of the class characteristic.
- x_i is the location of the specific subclass on the linear scale of the class characteristic.

It is also important to recognize that n_i/λ is a linear population "density". In the above examples, various n_i/λ are expressed as: soldiers per 1" of chest size, students per 4 units of test score, etc. ... and those "densities" are located at a specific place (x_i) in the overall range of chest sizes or test scores. To further illustrate that n_i/λ is a linear population density, let us assume that:

- The N soldiers (from one of the above examples) are all congregated on a large field.
- A line has been painted on the field and the line is also marked off in equal increments, with each increment or interval numbered from 31 to 55.
- The soldiers are instructed to stand within the numbered interval which corresponds to their chest size.

Assuming that the average chest size among all the soldiers is 43", a large number of soldiers will stand within the central interval marked 43. A lesser number of soldiers will stand within the adjacent intervals marked 42 and 44. An even lesser number of soldiers will stand within the further adjacent intervals marked 41 and 45. And so on, until only a very few soldiers will stand within the furthest intervals from the center, marked 31 and 55. Thus, the entire class of N soldiers will be distributed into subclasses of equal linear intervals as located by their chest sizes. The central interval will have the most soldiers per interval, or the highest linear "density". The intervals adjacent to the center interval will have lower linear densities and so on down the line in both directions from the center. The farthest intervals from the center will be sparsely populated and have the lowest linear densities.

Thus, the distribution of a large class population into smaller subclasses can be described as the distribution of linear population densities among defined intervals on the linear scale of a specific characteristic of the class. Such a distribution is very often presented in a graphical form as shown below:

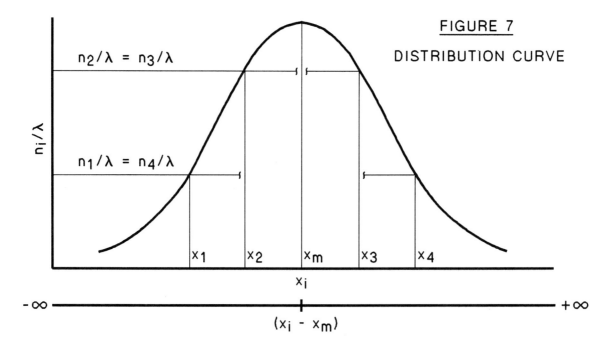

FIGURE 7

DISTRIBUTION CURVE

The distribution curve presented in Figure 7 might lead the reader to consider such a distribution as being an abstract mathematical function. In reality, it is a graphical representation of the connection of the mid-points at the bar tops of a linear distribution histogram as depicted in Figure 8 below:

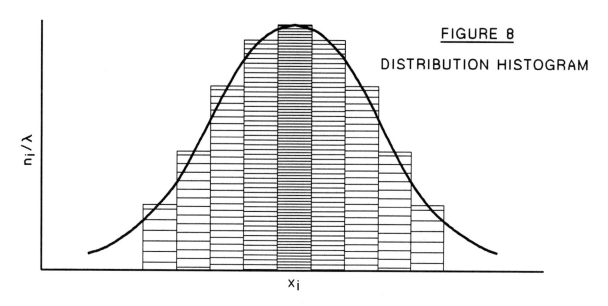

FIGURE 8

DISTRIBUTION HISTOGRAM

Each bar height represents a population density located along the linear x_i scale. The higher the density, the taller the bar. To stress the fact that a distribution curve is a graph of subclass population densities, the bars in Figure 8 are shaded to indicate the varying densities along the x_i scale. The center bar, representing the mean or average class characteristic, is the tallest bar and has the highest population density (n_i/λ) as indicated by the darkest shading. The other bars to the left and right of the center have progressively lower population densities as indicated by progressively shorter bars and lighter shading of the bars.

Many classes of statistical data can be grouped and presented to show how the amount of characteristic exhibited by individual members, or groups of members, deviate from the average characteristic of the overall class. When graphically plotted, many of them exhibit bell-shaped patterns very similar to those in Figures 7 and 8 where most of the population density along the x_i scale is distributed at and near to the average or mean value (x_m) of the overall class characteristic. In fact, the bell-shaped symmetrical distribution pattern is exhibited by so many types of statistical data that it is called the "normal distribution".

THE GAUSSIAN DISTRIBUTION

The bell-shaped distribution is also called the "Gaussian distribution". Its expression was derived by De Moivre, Gauss and Laplace in the seventeenth and eighteenth centuries:

(6) $\quad n_i/\lambda = (n_m/\lambda)\, e^{-(x_i - x_m)^2 / 2\sigma^2}$

Chapter 2: Gaussian Dispersion Equations

The standard deviation σ in equation (6) is a measure of the distribution of the class characteristic among its subclasses. The other terms in equation (6) are defined as:

x_m = the mean or arithmetic average value of the class characteristic amongst the total population

$$= \frac{\Sigma [n_i (1) (x_i/\lambda)]}{N} \quad \text{(for grouped data)}$$

$$= \frac{\Sigma (n_i x_i)}{N (\lambda)} \quad \text{(expressed in intervals)}^\dagger$$

$x_i - x_m$ = deviation in amount of class characteristic between members of subclass i and the mean subclass m (expressed in intervals)

σ = the standard deviation, or the root mean square deviation from x_m

$$= \sqrt{\frac{\Sigma [n_i (1) (x_i - x_m)^2 / \lambda^2]}{N - 1}}$$

$$= \sqrt{\frac{\Sigma [n_i (x_i - x_m)^2]}{\lambda^2 (N - 1)}} \quad \text{(expressed in intervals)}^{\dagger\dagger}$$

Subclasses may be located on the linear characteristic scale (see Figure 7) either by: their amount of characteristic (x_i); their specific deviation ($x_i - x_m$); or the number of standard deviations (σ's) equivalent to their specific deviation. In equation (6) as defined above, all of those linear characteristic scale values are expressed as intervals. If we deal only with intervals of 1 unit of dimension d per interval (i.e., λ of 1 \$/interval, 1"/interval, 1 m/interval, etc.), then we can dispense with the λ term and equation (6) becomes:

(6a) $\quad n_i = (n_m) e^{-(x_i - x_m)^2 / 2\sigma^2}$

where: $\quad x_m = \dfrac{\Sigma (n_i x_i)}{N} \quad$ in units of d

and: $\quad \sigma = \sqrt{\dfrac{\Sigma [n_i (x_i - x_m)^2]}{N - 1}} \quad$ in units of d

---

$\dagger \quad \left(\dfrac{\text{members}}{\text{interval}}\right)(1 \text{ interval})\left(\dfrac{d}{d / \text{interval}}\right) \div \text{members} = \text{intervals}$

$\dagger\dagger \quad \sqrt{\left(\dfrac{\text{members}}{\text{interval}}\right)(1 \text{ interval})\left(\dfrac{d^2}{d^2 / \text{interval}^2}\right) \div \text{members}} = \text{intervals}$

It is important to remember that equation (6a) is based on a λ of 1 unit of d per interval. Also, $(x_i - x_m)$ and σ must both have the same units of dimension so that the exponential term in the equation is dimensionless.

Given the σ and x_m of a class distribution, equation (6a) will provide the ratio of any subclass population to the mean subclass population (n_i/n_m). However, when dealing with stack gas dispersion, it is more useful to relate the population of a specific subclass (n_i) to the total class population. To do this, we can set $(x_i - x_m)$ equal to u for simplification, and then integrate equation (6a) to obtain the area under the curve ... which is the total class population N. Since that area is the summation of the rectangles having the height n_i and the width Δu:

$$N = \sum_{-\infty}^{\infty} (n_i \, \Delta u) = \int_{-\infty}^{\infty} n_i \, du = \int_{-\infty}^{\infty} n_m \, e^{-u^2/2\sigma^2} \, du$$

And the integration yields:

(7) $\quad N = n_m \, \sigma \, (2\pi)^{1/2}$

Dividing equation (6a) by equation (7), we obtain:

(8) $\quad \dfrac{n_i}{N} = \dfrac{e^{-(x_i - x_m)^2/2\sigma^2}}{\sigma (2\pi)^{1/2}}$

Equation (8) is the Gaussian distribution equation with the area under the curve equal to 1. To confirm that, we can integrate equation (8):

$$\sum_{-\infty}^{\infty} (n_i/N) \, \Delta u = \int_{-\infty}^{\infty} (n_i/N) \, du = \int_{-\infty}^{\infty} \dfrac{e^{-u^2/2\sigma^2}}{\sigma (2\pi)^{1/2}} \, du = \dfrac{\sigma (2\pi)^{1/2}}{\sigma (2\pi)^{1/2}} = 1$$

Assuming that the class distribution is indeed Gaussian, n_i/N is the fraction of the total class population which occurs within subclass i and it is called the "frequency" or "probability" of occurrence. For that reason, the area under the curve of equation (8) is said to have "unit frequency" or "unit probability".

Finally, equation (8) can be arranged in the form that has been widely utilized to develop stack gas dispersion models based upon the Gaussian distribution:

(9) $\quad n_i = \dfrac{N \, e^{-(x_i - x_m)^2/2\sigma^2}}{\sigma (2\pi)^{1/2}}$

Note that n_i in equation (9) is expressed in subclass members per unit of d, and therefore provides the linear population density of any subclass. Hence, equation (9) is often referred to as the "probability density function".

CONTINUOUS POINT-SOURCE PLUMES

The dispersion of emissions from a continuous point-source may be visualized as the conical plume shown in Figure 9. As the plume travels downwind, it may be further visualized as a series of disc-shaped increments which are diffusing and expanding in the vertical and the crosswind dimensions.

Assume that any one of the disc-shaped increments (a-b-c-d in Figure 9) is 1 meter wide in the downwind x-dimension and that:

- The point-source emissions (Q) are at a continuous mass-flow rate (in say grams/sec).

- The horizontal wind velocity (in say meters/sec) is constant and the mean wind direction is constant.

- Diffusion of emissions in the downwind direction is negligible relative to their transport by the wind. In other words, only vertical and crosswind diffusion occurs.

- There is no deposition or washout of emissions, no absorption of emissions by the ground or other physical bodies, and no chemical conversion of the emissions.

Thus, at any downwind distance x, the total plume volume (from the source to the point x) retains all of the emissions released during the time required for the wind to travel from the source to point x:

$Q(x/u)$ = weight of emissions in total plume (in grams) from source to point x

where: Q = point-source emissions, g/sec
u = horizontal wind velocity, m/sec
x = distance from point-source, m

And any crosswind increment, 1 meter of x in thickness, contains the emissions released in the time required for the wind to travel 1 meter:

Q/u = weight of emissions in 1 meter thick increment (in grams/meter)

Based on the generality that the <u>magnitude</u> of plume diffusion is a function of atmospheric turbulence and time, but that the actual <u>pattern</u> of diffusion is random (probabilistic), we can assume that the emissions in any crosswind increment of the plume will disperse in vertical and crosswind patterns that are essentially the same as given by the Gaussian distribution.[†]

Figure 9 depicts the Gaussian distribution of emission density in the vertical dimension of a crosswind increment, with a high density of emissions at the plume centerline interval and progressively decreasing to lower densities in a symmetrical pattern for the intervals above and below the centerline interval. A graph of those linear interval emission densities versus vertical distance above and below the centerline interval would have the same pattern as the Gaussian distributions in Figures 7 and 8.

[†] Slade[7] has an excellent discussion of the theory of molecular diffusion in stack gas plumes. However, in the final analysis, all that is needed is to assume Gaussian distribution.

Chapter 2: Gaussian Dispersion Equations 25

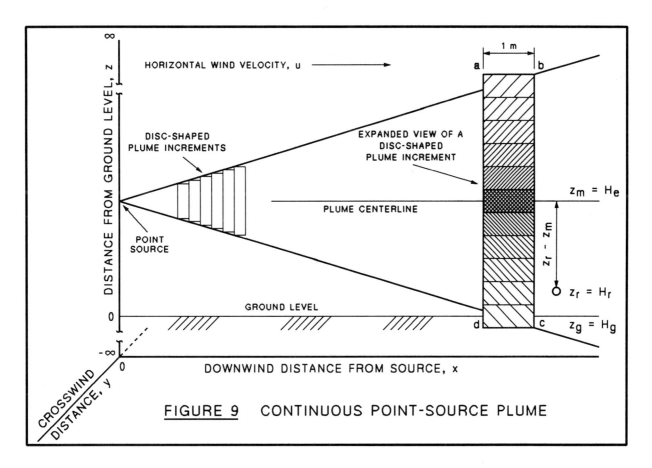

FIGURE 9   CONTINUOUS POINT-SOURCE PLUME

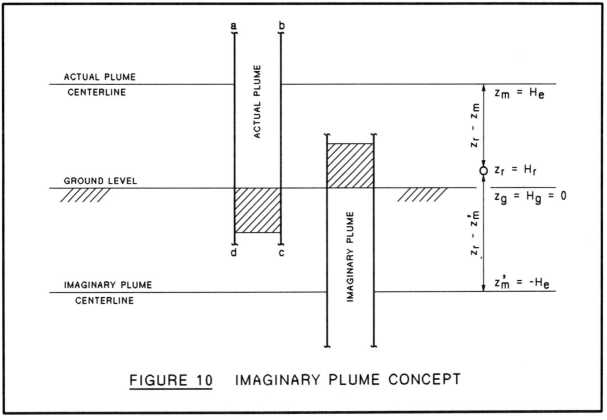

FIGURE 10   IMAGINARY PLUME CONCEPT

Chapter 2: Gaussian Dispersion Equations

For the moment, neglect the crosswind diffusion of the plume in Figure 9 and look only at the total emissions as <u>seen</u> by viewing the vertical dimension from a far distance, and hence seeing the <u>integrated</u> crosswind emissions in the 1 meter thick increment a-b-c-d ... without regard to any diffusion in the y-dimension. Applying the Gaussian distribution equation (9):

$$(9a) \quad n_r(x, z) = \frac{N e^{-(z_r - z_m)^2 / 2 \sigma_z^2}}{\sigma_z (2\pi)^{1/2}}$$

where:
- N = total grams of emissions in the entire 1 meter thick disc
- z_r = any receptor location (the small circle in Figure 9) in the z-dimension where we want to determine emission density
- z_m = location of the mean emission density (i.e., the plume centerline) in the z-dimension
- σ_z = vertical standard deviation of the emission densities, in meters (since λ = 1 m/interval)

The term n_r is that subpart of N distributed in the rth vertical interval and:

- N is the total emissions within the 1 meter thick increment, in grams/meter
- the exponential term is dimensionless
- σ_z is in meters

Thus, equation (9a) is dimensionally consistent and n_r is obtained as a rectangular emission density in grams per square meter (g/m^2). Therefore:

$n_r(x, z)$ is the integrated crosswind emission density, in g/m^2, seen at the receptor located at z_r when viewing the x-z plane

Equation (9a) <u>assumes</u> that there is no barrier to upward or downward diffusion in the vertical dimension. For that reason, Figure 9 depicts finite disc-shaped increments and finite plume boundaries. However, a true Gaussian distribution would extend infinitely above and below the mean emission density at the plume centerline. Thus, we must make another assumption: that any emissions reaching the ground (which is a barrier) are totally reflected upward since we have already assumed no deposition, washout or absorption. To determine n_r at any ground-level or above-ground receptor, how do we account for the emissions which are reflected upward?

We do so by assuming that an imaginary plume exists beneath ground-level which is identical to the actual above-ground plume, and which is at the same distance from ground-level as is the actual plume (see Figure 10) and at the same downwind distance from the source.

As shown in Figure 10, the shaded area of the actual plume increment (representing the emissions which cannot penetrate the ground and are reflected upward) is replaced by the shaded area of the imaginary plume increment. In other words, the shaded portion of the imaginary plume distribution is added to the unshaded distribution of the actual plume.

Thus, the Gaussian distribution in the vertical or z-dimension, including upward reflection from the ground, becomes:

$$(10) \quad n_r(x, z) = \frac{N e^{-(z_r - z_m)^2 / 2\sigma_z^2}}{\sigma_z (2\pi)^{1/2}} + \frac{N e^{-(z_r - z'_m)^2 / 2\sigma_z^2}}{\sigma_z (2\pi)^{1/2}}$$

Now we can make these substitutions in equation (10):

$$\begin{aligned} N &= Q/u & &\text{(see pages 24 and 26)} \\ z_r - z_m &= H_r - H_e & &\text{(see Figure 10)} \\ z_r - z'_m &= H_r - (-H_e) & &\text{(see Figure 10)} \\ &= H_r + (H_e) \end{aligned}$$

where: H_e = height of plume centerline above ground, m
H_r = height of receptor above ground, m

And we obtain:

$$(11) \quad n_r(x, z) = \frac{Q}{u \sigma_z (2\pi)^{1/2}} \left[e^{-(H_r - H_e)^2 / 2\sigma_z^2} + e^{-(H_r + H_e)^2 / 2\sigma_z^2} \right]$$

Now we can include the <u>crosswind</u> Gaussian distribution of $n_r(x, z)$ in the y-dimension:

$$(12) \quad n_r(x, y, z) = \frac{n_r(x, z) \, e^{-(y - y_m)^2 / 2\sigma_y^2}}{\sigma_y (2\pi)^{1/2}}$$

There is no diffusion barrier in the crosswind dimension and hence no need for another reflection term in equation (12).

Finally, we can make these substitutions into equation (12):

$$\begin{aligned} C &= n_r(x, y, z) \text{ to conform with convention}^\dagger \\ z_r &= H_r \text{ to conform with convention}^\dagger \\ n_r(x, z) &= \text{the right-hand side of equation (11)} \\ y_m &= 0 \text{ for the location of the mean emission density at the plume centerline in the crosswind or y-dimension} \\ y &= \text{distance from the receptor to the plume centerline in the crosswind or y-dimension} \end{aligned}$$

---

† Convention means the usual form in which these terms are expressed in the literature.

28   Chapter 2: Gaussian Dispersion Equations

After making the above substitutions into equation (12), we obtain the well-known **GENERALIZED GAUSSIAN DISPERSION EQUATION FOR A CONTINUOUS POINT-SOURCE PLUME:**

(13) $$C = \frac{Q}{u\,\sigma_z\,\sigma_y\,2\pi}\,e^{-y^2/2\sigma_y^2}\left[e^{-(z_r-H_e)^2/2\sigma_z^2} + e^{-(z_r+H_e)^2/2\sigma_z^2}\right]$$

where:
- C = concentration of emissions, g/m³, at any receptor located at:
  - x meters downwind
  - y meters crosswind from the centerline
  - z_r meters above ground
- Q = source emission rate, g/sec
- u = horizontal wind velocity, m/sec
- H_e = plume centerline height above ground, m
- σ_z = vertical standard deviation of the emission distribution, m
- σ_y = horizontal standard deviation of the emission distribution, m

Equation (13) is valid only within these summarized constraints:

- Vertical and crosswind diffusion occur according to Gaussian distribution.

- Downwind diffusion is negligible compared to downwind transport.

- The emissions rate, Q, is continuous and constant.

- The horizontal wind velocity and the mean wind direction are constant.

- There is no deposition, washout, chemical conversion or absorption of emissions, and any emissions diffusing to the ground are reflected back into the plume (i.e., all emissions are totally conserved within the plume).

- There is no upper barrier to vertical diffusion and there is no crosswind diffusion barrier.

- Emissions reflected upward from the ground are distributed vertically as if released from an imaginary plume beneath the ground and are additive to the actual plume distribution.

- The use of σ_z and σ_y as constants at a given downwind distance, and the assumption of an expanding conical plume (see Figure 9), implicitly require homogeneous turbulence throughout the x, y and z-dimensions of the plume.

The above summary makes it fairly obvious that the plume dispersion equation (13) for a continuous point-source depends upon the validity of a good many assumptions. In addition, valid and accurate values of σ_z and σ_y are required as a function of downwind distance. Similarly, the height of the plume centerline (H_e) must be realistically determined as a function of downwind distance. As long as all of the constraints and assumptions are kept in mind, equation (13) is the generalized Gaussian dispersion equation for a continuous point-source plume from which other more specific plume dispersion equations may be derived.

In terms of the environmental impact of plume components such as SO_2, NO_x, and others, the primary concern is usually with their ground-level concentrations. For that case, the receptor is at $z_r = 0$ and the **GROUND-LEVEL CENTERLINE AND CROSSWIND CONCENTRATIONS** are obtained by reducing equation (13) to :

$$(14) \quad C = \frac{Q}{u\,\sigma_z\,\sigma_y\,\pi}\; e^{-y^2/2\sigma_y^2}\; e^{-H_e^2/2\sigma_z^2}$$

or the equivalent form of:

$$(14a) \quad C = \frac{Q}{u\,\sigma_z\,\sigma_y\,\pi}\; e^{-(y^2/2\sigma_y^2)-(H_e^2/2\sigma_z^2)}$$

Since, by definition, a Gaussian distribution is asymmetrical about its mean interval of maximum density, the crosswind concentrations are always lower than the centerline concentration. Thus, if the primary concern is with the environmental impact of the maximum ground-level concentrations, the receptor is at $z_r = 0$ and $y = 0$. For that case, the **GROUND-LEVEL CENTERLINE CONCENTRATIONS** are obtained by reducing equation (13) to:

$$(15) \quad C = \frac{Q}{u\,\sigma_z\,\sigma_y\,\pi}\; e^{-H_e^2/2\sigma_z^2}$$

Finally, if the interest is in **GROUND-LEVEL CENTERLINE CONCENTRATIONS FROM GROUND-LEVEL PLUMES** (receptor at $z_r = 0$, $y = 0$ and $H_e = 0$), equation (13) reduces to:

$$(16) \quad C = \frac{Q}{u\,\sigma_z\,\sigma_y\,\pi}$$

VERTICAL CONCENTRATION PROFILES

Although the primary environmental concern with stack gas plumes is with their ground-level concentrations, their vertical concentration profiles may also be of concern. For example, in a petroleum refinery or petrochemical plant, there may be many working platforms at elevations of 20-40 meters and at distances of 100 to 1500 meters from the plant's source emission points. Operating or maintenance personnel working on those platforms will be exposed to the pollutant concentrations within the dispersing plumes existing at those elevations whenever the platforms are downwind of the emission source points.

In any event, a discussion of the vertical concentration profile of a continuous point-source plume will provide a better understanding of how stack gas plumes disperse. Since the maximum concentrations exist in a vertical plane through the plume centerline, we will let $y = 0$ and eliminate the crosswind exponential term from the generalized Gaussian dispersion equation (8) to obtain the **CENTERLINE VERTICAL CONCENTRATIONS** from:

$$(17) \quad C = \frac{Q}{u\,\sigma_z\,\sigma_y\,2\pi}\left[e^{-(z_r-H_e)^2/2\sigma_z^2} + e^{-(z_r+H_e)^2/2\sigma_z^2}\right]$$

Chapter 2: Gaussian Dispersion Equations

Figure 11 illustrates some vertical concentration profiles, obtained from equation (17), for various emission heights at a fixed downwind distance from the emission point-source of one kilometer based upon these specifics:

Emission heights (H_e)	50-200 m
Receptor elevations (z_r)	0-500 m
Wind velocity (u)	4 m/sec
SO_2 emission rate (Q)	151×10^6 µg/sec[†]
Downwind distance (x) from source	1 km
At x of 1 km and Pasquill stability class B:	
σ_z	110 m[††]
σ_y	157 m[††]

As seen in Figure 11, when a stack gas plume has travelled one kilometer downwind, the plume's vertical concentration profile is no longer symmetrical and no longer resembles the Gaussian distribution pattern depicted in Figures 7 and 8. The plume has grown to the extent that plume components reaching the ground and being reflected upward have considerably altered the vertical dispersion pattern. As also seen in Figure 11, the effect of upward reflection from the ground is more pronounced for lower emission heights (H_e). In fact, the reflection effect is so pronounced for the 50 meter emission height that the plume's emission concentration at ground-level (where $z_r = 0$) is higher than at the plume's centerline height of 50 meters. The same is true for the 115 meter emission height.

Figure 12 presents more vertical concentration profiles, also obtained from equation (17), for a fixed emission height (H_e) of 150 meters at various downwind distances based upon these specifics:

Emission height (H_e)			150 m
Receptor elevations (z_r)			0-600 m
Wind velocity (u)			4 m/sec
Downwind distance (x)	0.5 km	1.0 km	2.0 km
At distance x and Pasquill stability class B:			
σ_z	51 m	110 m	232 m
σ_y	83 m	157 m	193 m

As seen in Figure 12, the vertical concentration profile at one-half kilometer downwind resembles the Gaussian distribution pattern in Figures 7 and 8. The plume has not yet expanded to the extent that ground reflection has significantly affected the vertical dispersion pattern. However, at one kilometer downwind, the effect of reflection on the profile is quite significant and similar to that seen in Figure 11. At two kilometers downwind, the reflection effect is so pronounced that the plume's emission concentration at ground-level is higher than at the plume's centerline height.

Figure 13 presents vertical concentration profiles based upon specifics similar to those used

---

[†] $(151 \times 10^6$ µg/sec$)(1$ g/10^6 µg$)(3600$ sec/hr$)(1$ lb/454 g$) = 1197$ lbs/hr

[††] Determination of σ values as functions of downwind distance and Pasquill stability class is discussed in Chapter 3.

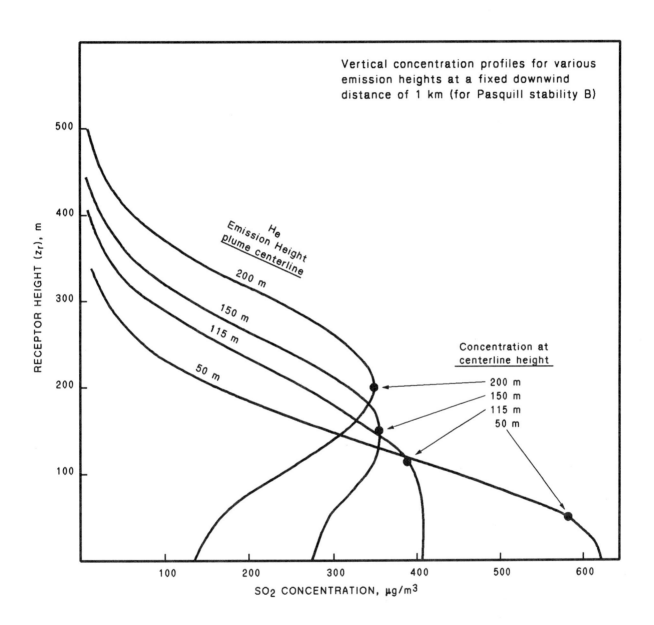

FIGURE 11

VERTICAL CONCENTRATION PROFILE AT PLUME CENTERLINE

Chapter 2: Gaussian Dispersion Equations

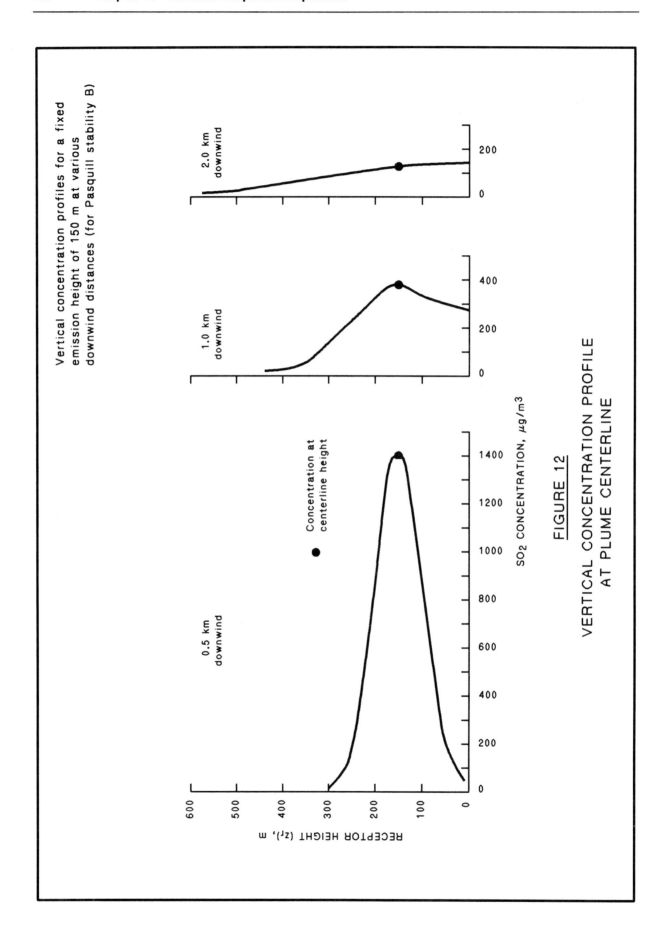

FIGURE 12
VERTICAL CONCENTRATION PROFILE
AT PLUME CENTERLINE

Chapter 2: Gaussian Dispersion Equations 33

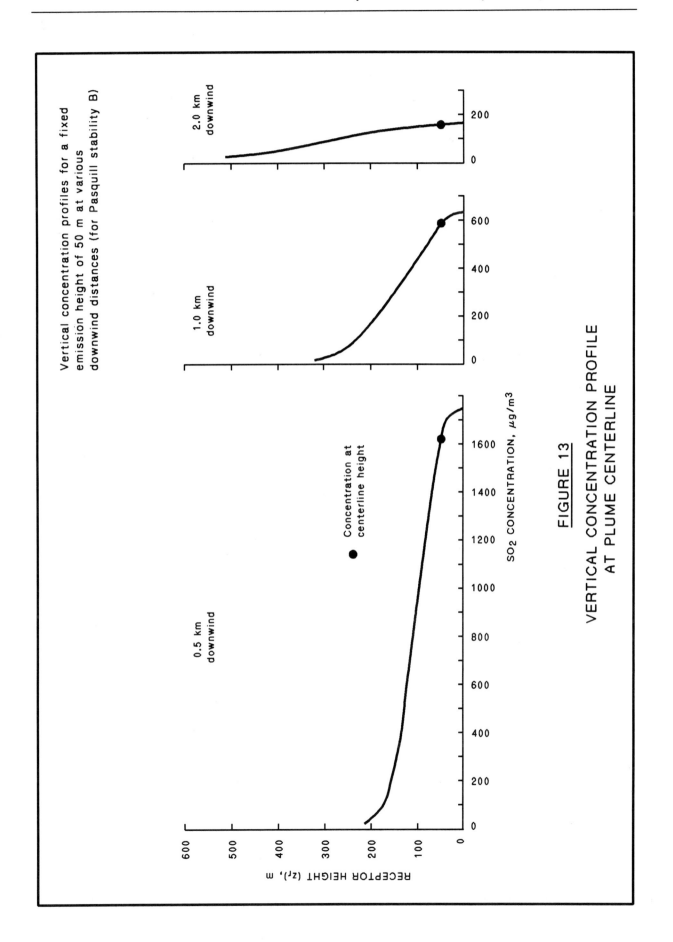

FIGURE 13
VERTICAL CONCENTRATION PROFILE
AT PLUME CENTERLINE

to obtain the profiles in Figure 12 except that the emission height (H_e) was fixed at a lower level of 50 meters. As seen in Figure 13, the effect of ground reflection is very pronounced at all three downwind distances of one-half, one, and two kilometers.

In summary, Figures 12 and 13 both exhibit vertical concentration profiles with these same characteristics:

- The plume's vertical dispersion pattern expands vertically as it travels downwind.

- The vertical concentration profile "flattens out" as it travels downwind, for two reasons:

    (a) The vertical standard deviation of the distribution (σ_z) increases with increasing downwind distance.

    (b) The reflection effect becomes more pronounced with increasing downwind distance.

- Although the plume's vertical concentration profile may initially have a maximum node at the plume centerline, the node is eventually eliminated as the plume travels downwind.

- Vertical concentration profiles deriving from plumes with lower emission heights (plume centerline heights) are influenced more by the reflection effect than are profiles deriving from plumes with higher emission heights.

The above characteristics of vertical concentration profiles are all logical consequences of the derivation of the generalized Gaussian dispersion equation (13).

CROSSWIND CONCENTRATION PROFILES

Crosswind concentration profiles at various receptor elevations are also of interest in terms of their environmental impact upon elevated inplant working platforms, and in terms of better understanding the dispersion characteristics of stack gas plumes.

Since the maximum crosswind concentrations exist in a horizontal plane located at the plume centerline height (where $z_r = H_e$), the following equation (18) for the **CROSSWIND CONCENTRATIONS AT THE PLUME CENTERLINE HEIGHT** can be obtained from the generalized Gaussian distribution equation (13):

$$(18) \quad C = \frac{Q}{u\,\sigma_z\,\sigma_y\,2\pi}\, e^{-y^2/2\sigma_y^2} \left[1 + e^{-2H_e^2/2\sigma_z^2} \right]$$

The exponential term within the brackets represents the upward reflection of plume components from the ground. Since there is no crosswind barrier to diffusion, the ground reflection has a uniform effect over the entire y-dimension of the crosswind concentration profiles. Thus, a crosswind profile is symmetrical around its central maximum node.

Figure 14 illustrates the effect of emission height (i.e., plume centerline height) upon crosswind concentration profiles at the plume centerline height for various distances downwind from the plume source. Figure 14 is based upon the same specifics as were

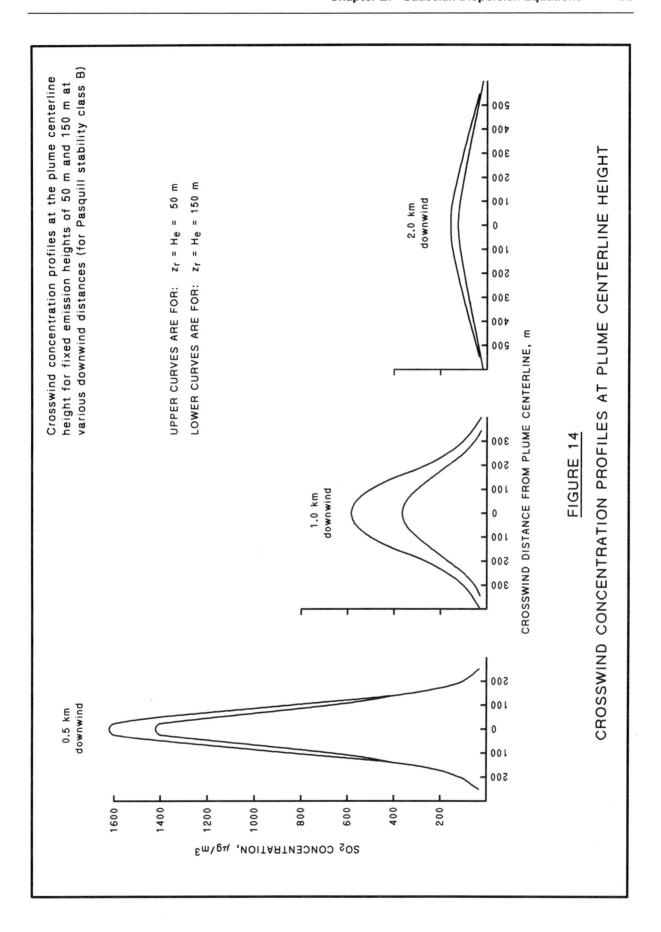

FIGURE 14

CROSSWIND CONCENTRATION PROFILES AT PLUME CENTERLINE HEIGHT

Figures 12 and 13, and it exhibits these characteristics of crosswind dispersion profiles:

- The plume's crosswind dispersion pattern expands horizontally as it travels downwind from the emission source.

- The crosswind concentration profile "flattens out" as the plume travels downwind, because the horizontal standard deviation of the plume's dispersion (σ_y) increases with increasing downwind distance.

- The crosswind concentration profile maintains Gaussian symmetry around the central maximum node of the profile.

- Lower emission heights (H_e) result in higher crosswind concentrations at the plume centerline heights, because of the higher effect of ground reflection for lower emission heights.

It should be noted that the six profiles in Figure 14 are simply the crosswind concentration profiles for the six plume centerline points denoted by the small black circles in Figures 12 and 13.

Figure 15 shows how the crosswind concentration profiles, for a fixed emission height (H_e) of 150 meters, vary from the plume centerline height down to ground level:

0.5 km downwind -- the crosswind concentrations at the plume centerline are much higher than the crosswind concentrations at the ground, and Figure 15 includes the crosswind concentration profiles at intermediate heights of 100 meters and 75 meters.

1.0 km downwind -- the crosswind concentrations at the plume centerline are only slightly higher than the crosswind concentrations at the ground.

2.0 km downwind -- the crosswind concentrations at the plume centerline are actually somewhat <u>lower</u> than the crosswind concentrations at the ground.

In other words, the crosswind concentration profiles in Figure 15 generally exhibit the same ground reflection effects as the vertical concentration profiles in Figure 12 exhibited for the same emission height of 150 meters. This emphasizes the fact that the ground's reflection of dispersing plume components eventually eliminates the centerline maximum node from the vertical concentration profiles but not from the crosswind concentration profiles.

As discussed in a later section of this chapter as well as in Chapter 8, upward reflection of plume components which have diffused to the ground is also an important factor when upward diffusion becomes limited by the presence of an inversion aloft and a dispersing stack gas plume becomes "trapped" beneath the inversion aloft.

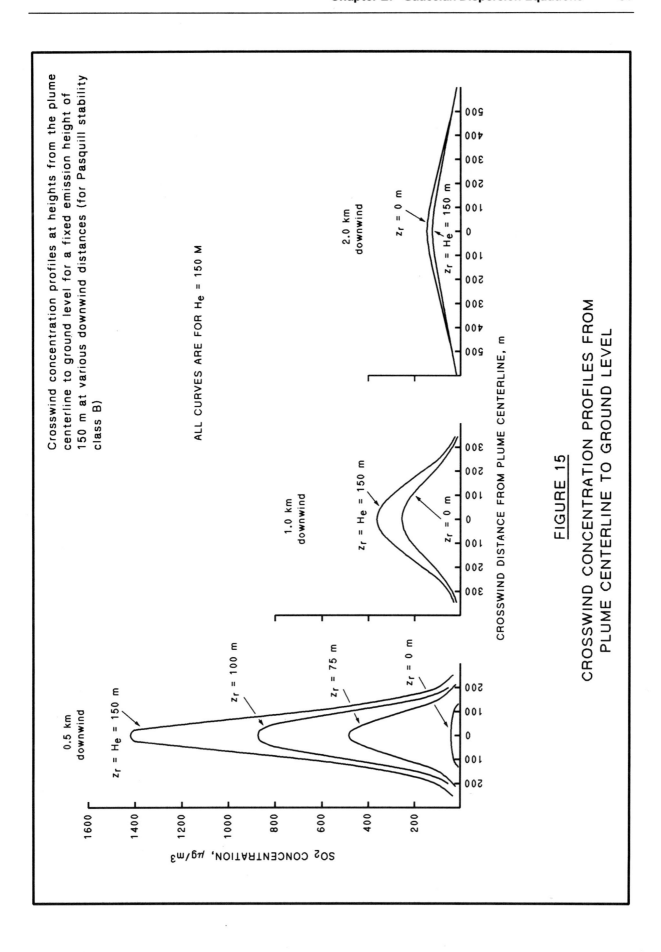

FIGURE 15

CROSSWIND CONCENTRATION PROFILES FROM PLUME CENTERLINE TO GROUND LEVEL

DISCUSSION OF INPUT PARAMETERS

Thus far in this book, we have dealt with the use of ambient temperature gradients in classifying atmospheric turbulence or stability. We have also discussed inversion layers and mixing limits or mixing heights. We next examined statistical distributions in general and the Gaussian distribution in particular.

We then derived the generalized Gaussian dispersion equation for continuous point-source stack gas plumes, and examined the vertical and crosswind dispersion characteristics of such plumes.

As yet, we have not discussed the determination or selection of the input parameters required by the Gaussian dispersion equation. Those determinations and selections will be dealt with in the subsequent chapters of this book. However, at this point, it would be useful to briefly summarize just what information must yet be examined:

Plume Centerline Height Or Emissions Height (H_e)

H_e is often referred to as the "effective stack height", which should not be confused with the actual height of the emission source. The effective stack height or emissions height is greater than the actual source height by the amount that the plume rises after it issues from the source stack or vent. Thus,

(19) H_e = source height + plume rise

Since the Gaussian dispersion equation requires the emissions height (H_e), we must learn how to calculate the plume rise so that we can obtain the emissions height.

We must also learn that *H_e is not a constant* across the entire x-dimension. In other words, H_e is not "the" emissions height or "the" effective stack height, because the plume rise component of H_e varies considerably as the plume travels downwind from the source.

Receptor Location

The receptor is the point at which we wish to calculate an emission concentration. It is located by its height above ground-level (z_r), and by its crosswind distance (y) from the plume's vertical centerline plane.

Although, the downwind distance from the emission source to the receptor (x) does not appear in the Gaussian dispersion equation, it is one of the factors in determining the plume rise as well as the σ_z and σ_y values. Thus, it is a required input parameter or specification.

The receptor location in terms of x, y and z_r requires no further elaboration beyond the recognition that it is an input parameter or specification.

Source Emission Rate (Q)

The source emission rate is also an input parameter or specification. Thus far, we have defined the source emission rate (Q) as having units of g/sec or µg/sec from which it followed that emission concentrations (C) obtained from the Gaussian dispersion equation have units of g/m^3 or $µg/m^3$. However, any set of consistent units may be used. For example, if we define

> E = any emission units (g, μg, ft^3, m^3, etc.)

and

> Q = E/sec

then we obtain the emissions concentration C in units of E/m^3.

To avoid any confusion, all of the other terms in the Gaussian dispersion equation (i.e., y, z_r, H_e, u, σ_z and σ_y) are best kept in the units as defined on page 28 in this chapter.

Horizontal Wind Velocity (u)

The horizontal wind velocity is an input parameter or specification for which we must yet determine whether to use the wind velocity at: ground-level, or at the source height, or at the effective stack height (H_e). We must also examine how wind velocities measured at ground-level are correlated to velocities at the source height or at the effective stack height.

The Dispersion Coefficients (σ_z and σ_y)

The derivation of the Gaussian dispersion equation requires that σ_z and σ_y are constants throughout the vertical z-dimension and the horizontal y-dimension.[†] We have yet to discuss the fact that the commonly used values of σ_z and σ_y are functions of: the downwind distance x from the source to the receptor, and the atmospheric turbulence as defined by the Pasquill stability class.

As mentioned earlier, derivation of the Gaussian dispersion equation requires homogeneous atmospheric turbulence throughout the x, y and z-dimensions of the dispersing plume.[†]

The reader should note that, in all the subsequent sections of this book, σ_z and σ_y will be referred to as "dispersion coefficients" rather than "the vertical and horizontal standard deviations of the emission distribution". That terminology is adopted merely for the purpose of simplification.

[†] Derivation of the Gaussian dispersion equation (13) involves integrations which require σ_z and σ_y to be constants. Since σ_z and σ_y are functions of atmospheric turbulence, it follows that the derivation also requires constant (i.e., homogeneous) turbulence in the z and y-dimensions. Although the derivation of equation (13) does not <u>explicitly</u> require constant or homogeneous turbulence in the x-dimension, it is <u>implicit</u> in the assumption of an expanding conical plume as illustrated in Figure 9. As the plume travels through any downwind distance (x) of turbulence, it expands to some size in the z and y-dimensions. If it were then to travel through another increment of downwind distance with lower turbulence (and hence lower σ values), the Gaussian dispersion equation might predict shrinkage of the plume dimensions. Obviously, this cannot occur and is inconsistent with the assumption of an expanding plume.

It should be noted that "conical" may not be an accurate description of the idealized Gaussian plume because it does not expand uniformly in the z and y-dimensions (i.e., σ_z and σ_y are not usually equal).

Chapter 2: Gaussian Dispersion Equations

TYPICAL PLUME BEHAVIOR MODES

The generalized Gaussian dispersion equation (13) was derived to predict the dispersion behavior of the idealized conical plume in Figure 9. However, all "real world" plumes do not behave in the idealized manner. Figure 16 illustrates six types of plumes commonly described in the literature.[2,7] There is a small diagram of altitude versus temperature at the left of each plume type in Figure 16. The diagrams characterize the atmospheric conditions conducive to the formation of each plume type, in terms of the relation between the existing ambient temperature gradient (solid lines) and the dry adiabatic lapse rate (dashed lines).

Looping Plume

This plume behavior usually occurs during unstable, super-adiabatic atmospheric conditions as indicated by the ambient temperature gradient and the dry adiabatic lapse rate in the small diagram at the left of the looping plume in Figure 16. As discussed in Chapter 1, such conditions are characterized by a high degree of vertical turbulence which causes the plume to fluctuate or "loop" in the vertical plane. A mean centerline through the looping plume might resemble the centerline of the idealized conical plume. Hence, the generalized Gaussian dispersion equation might predict the mean behavior of a looping plume, but would not predict the localized, high ground-level concentration of plume components where the plume is brought to ground by its fluctuations.

Typically, the atmospheric conditions conducive to forming a looping plume will occur on warm days with clear skies with little wind (see Table 1, Pasquill stability classes A or B).

Coning Plume

This plume behavior usually occurs during near neutral atmospheric conditions with ambient temperature gradients near to the dry adiabatic lapse rate, as indicated in the small diagram at the left of the coning plume in Figure 16. Obviously, the generalized Gaussian dispersion equation is applicable to the predicting of the dispersion behavior of coning plumes since the derivation of the generalized Gaussian equation is based upon an idealized coning plume.

Typically, the atmospheric conditions conducive to forming a coning plume will occur on windy, cloudy days or windy nights (see Table 1, Pasquill stability class D).

Fanning Plume

This plume behavior occurs when a plume is imbedded within an inversion layer, either a surface inversion layer or an inversion aloft layer, as indicated by the two ambient temperature gradients in the small diagram at the left of the fanning plume in Figure 16. The very stable conditions within the inversion layer inhibit vertical turbulence, and the plume exhibits very little expansion in the vertical plane. However, the plume "fans out" in the crosswind plane because σ_z is quite small relative to σ_y.

For a fanning plume imbedded within a surface inversion, the generalized Gaussian dispersion equation is applied by using σ values appropriate to the stable conditions within the inversion layer. However, that probably overstates the plume's ground-level concentrations because the increased dispersion in the more turbulent layer above the inversion is not accounted for by using σ values for the stable inversion layer. This is an example of dispersion during atmospheric conditions which are not homogeneous throughout the vertical dimension.

Chapter 2: Gaussian Dispersion Equations 41

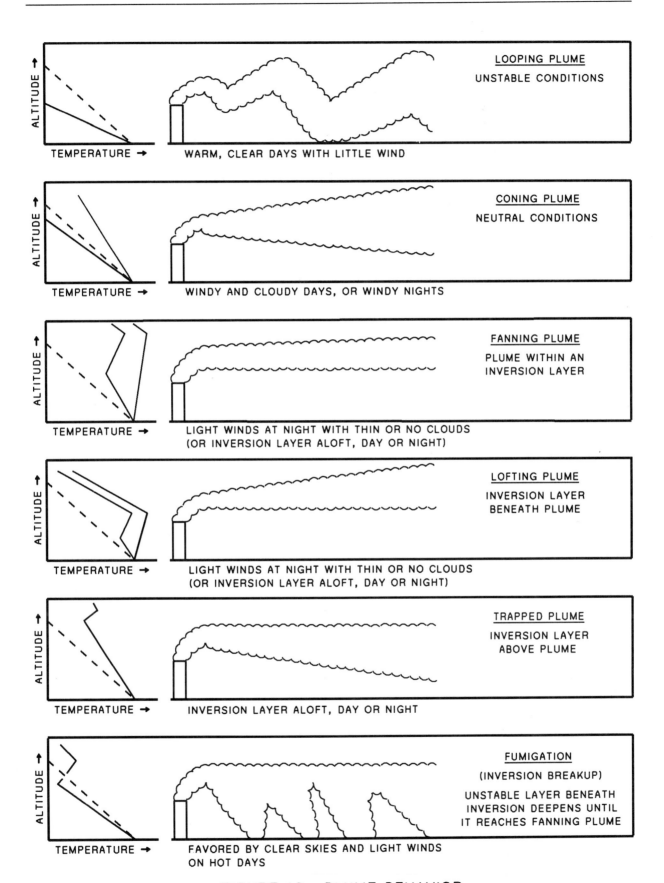

FIGURE 16   PLUME BEHAVIOR

For a fanning plume imbedded within an inversion aloft, downward dispersion passes through two layers of different atmospheric stability: (a) the inversion layer itself and (b) the surface layer beneath, which may be either super-adiabatic, neutral or sub-adiabatic (see the discussion of Figure 3 in Chapter 1). This makes it difficult to select appropriate σ values for the generalized Gaussian dispersion equation. The same problem exists for upward dispersion through the atmospheric layer above the inversion aloft. To some extent, increased dispersion through the more turbulent layers above and below the inversion may offset each other, but that possibility is complicated by the effect of ground reflection on the dispersion pattern. All in all, it is most difficult to decide whether the ground-level concentrations resulting from a fanning plume are over or under-stated by using σ values for the stable conditions within the inversion layer. There are too many permutations of layer heights, layer depths, and relative positions of the plume within the inversion layer.

In any event, a fanning plume imbedded in an inversion layer exhibits very little vertical dispersion and, therefore, the effect of upward and downward dispersion through the adjacent layers of turbulence may not be significant.

The atmospheric conditions conducive to forming surface inversions will usually occur during nights with light winds and clear skies (see Table 1, Pasquill stability class F). Inversions aloft may form by the day-time break-up of the previous night's surface inversion, or they may be "frontal" or "subsidence" inversions which occur either at night or during the day.

Lofting Plume

This plume behavior occurs when the plume is above an inversion layer and downward dispersion is blocked by the stable inversion layer beneath the plume. The lofting plume may be above a surface inversion or an inversion aloft, as indicated by the two ambient temperature gradients in the small diagram at the left of the lofting plume in Figure 16.

If the downward dispersion is assumed to be reflected upward by the inversion layer (rather than being absorbed and trapped within the inversion), then the generalized Gaussian dispersion equation could be applied to lofting plumes by re-defining the vertical z-dimension coordinates. However, since the primary concern is usually with ground-level concentrations, there may be little purpose in re-defining the z-dimension coordinates because dispersion to the ground from lofting plumes is essentially nil.

Typically, the atmospheric conditions conducive to forming a lofting plume are the same as for a fanning plume, since both the lofting and fanning plumes require the existence of a surface inversion or an inversion aloft.

Trapped Plume

This plume behavior occurs when the plume is below an inversion aloft and thus upward dispersion is blocked by the inversion layer above the plume, as indicated by the ambient temperature gradient in the small diagram at the left of the trapped plume in Figure 16.

Various methods have been suggested for predicting the dispersion of trapped plumes[2,8,9,10] and they will be examined in more detail in Chapter 8 of this book. However, two of those methods are summarized briefly at this point.

One method assumes that the generalized Gaussian dispersion equation (13) is applicable from the emission source to some downwind distance, x_L, at which the mixing height (which is the distance from the ground to the base of an inversion aloft) is equal to the plume centerline height plus 2.15 σ_z :

(20) Mixing height, $L = H_e + 2.15\ \sigma_z$

It is also assumed that, at distance x_L, the plume concentrations impinging on the base of the inversion aloft amount to 10 percent of the plume centerline concentrations and plume trapping begins to significantly affect the vertical dispersion pattern of the plume at that point. It is further assumed that, at twice that distance (i.e., 2 x_L), the plume's vertical concentrations are no longer dispersed in a Gaussian pattern and instead are uniformly dispersed within the mixing height L. In other words, at downwind distance 2 x_L and beyond, crosswind dispersion is the only factor affecting plume concentrations. The vertical distribution frequency becomes constant and equal to 1/L. Therefore, the dispersion equation for a trapped plume may be taken as:

(21) $$C = \frac{Q}{u L \sigma_y (2\pi)^{1/2}} e^{-y^2/2\sigma_y^2}$$

Concentrations between the distances x_L and 2 x_L are obtained by log-log interpolation between the concentrations at x_L and 2 x_L. This method for predicting the dispersion of trapped plumes was suggested by Pasquill[8] in 1961 and expanded upon by Turner[9] in 1970. However, Turner defined the mixing height L at x_L as being 2.15 σ_z ... which is incorrect for elevated plumes, and which is correctly defined by equation (20).[†]

Another commonly suggested method[10] accounts for the effect of multiple reflections between the ground and the base of the inversion aloft. If the upward dispersion of the trapped plume is reflected downward by the inversion base (rather than being absorbed within the inversion), this method is probably more rigorous than Pasquill and Turner's method. Chapter 8 of this book is devoted to developing this method for predicting the dispersion of trapped plumes by accounting for the effect of multiple reflections between the ground and the base of the inversion aloft.

Typically, the atmospheric condition conducive to forming a trapped plume is the existence of an inversion aloft which may result from the day-time break-up of the previous night's surface inversion, or may be a "frontal" or "subsidence" inversion occurring either day or night.

Fumigation

This plume behavior occurs when the plume, which was imbedded within a surface inversion as a fanning plume, is broken up by rising turbulence when day-time heating of the ground breaks up the inversion. The fumigation results in high ground-level concentrations of short duration where the broken plume impinges upon the ground at various distances from the emission source.

---

† Even corrected as in equation (20), Turner's method is best used only for small ratios of H_e to L (see Chapter 8).

Typically, the atmospheric conditions conducive to forming a fumigation occur on hot days with clear skies and light winds (see Table 1, Pasquill stability class A or B). A method of predicting dispersion from fumigated plumes is developed in detail in Chapter 9 of this book.

BOUNDARY-TO-CENTERLINE CONCENTRATION RATIOS

As mentioned earlier in this chapter, Gaussian distributions extend to infinity on either side of their central mean intervals. However, the emissions concentration at a selected boundary may be determined for any Gaussian plume dispersion by defining the distance between the plume centerline and the selected boundary. For any defined distance between the mean interval and a selected boundary interval (the ith interval), the boundary-to-centerline emissions concentration ratio can be determined by using equation (6a):

(6a) $$\frac{n_i}{n_m} = e^{-(x_i - x_m)^2 / 2\sigma^2}$$

and re-arranging:

$$(x_i - x_m) = \sigma [-2 \ln(n_i / n_m)]^{1/2}$$

and in terms of the generalized Gaussian dispersion equation:

(22) $$(z_r - H_e) = \sigma_z [-2 \ln(C_r / C_e)]^{1/2}$$

where: C_r = emission concentration at receptor height z_r
C_e = emission concentration at plume centerline height H_e
ln = logarithm to the base e (i.e., natural logarithm)

Equation (22) provides the distances from the plume centerline corresponding to various boundary-to-centerline emissions concentration ratios. For example:

C_r/C_e	$(z_r - H_e)$
0.01	3.03 σ_z
0.05	2.45 σ_z
0.10	2.15 σ_z

Thus, at a vertical distance of 3.03 σ_z from the plume centerline, the boundary concentration of emissions will be 1 percent of the centerline concentration. And at a vertical distance of 2.15 σ_z from the plume centerline, the boundary concentration will be 10 percent of the centerline concentration. The latter boundary at 2.15 σ_z is frequently referred to in the literature for various reasons, and we have already discussed its use in defining the point x_L at which the dispersion pattern of a trapped plume is assumed to become non-Gaussian.

Chapter 3

DISPERSION COEFFICIENTS

THE HISTORICAL DEVELOPMENT OF DISPERSION COEFFICIENTS

One of the earliest stack gas plume dispersion equations was derived by Bosanquet and Pearson[11] in 1936. The equation they developed for the **GROUND-LEVEL CONCENTRATIONS OF A CONTINUOUS POINT SOURCE PLUME** was:

$$(23) \quad C = \frac{Q}{u p q x^2 (2\pi)^{1/2}} \, e^{-y^2/2q^2 x^2} \, e^{-H_e/px}$$

Bosanquet and Pearson's derivation of equation (23) did not directly assume Gaussian distribution nor did it include the effect of ground reflections. Bosanquet and Pearson concluded that their dimensionless dispersion constants p and q had average values of about 0.05 and 0.08, respectively.

Sir Graham Sutton[12] derived a Gaussian distribution equation in 1947, from which this equation can be obtained for the **GROUND-LEVEL CONCENTRATIONS OF A CONTINUOUS POINT SOURCE PLUME:**

$$(24) \quad C = \frac{Q}{u \, \tfrac{1}{2} \, C_y C_z \pi x^{2-n}} \, e^{-y^2/C_y^2 x^{2-n}} \, e^{-H_e^2/C_z^2 x^{2-n}}$$

Sutton's equation included the effect of ground reflection and the assumption of Gaussian distribution for the vertical and crosswind dispersion of the plume.

Sutton's equation is very similar to equation (14) for the **GROUND-LEVEL CONCENTRATIONS OF A CONTINUOUS POINT SOURCE PLUME** which was derived in Chapter 2 from the generalized distribution equation (13):

$$(14) \quad C = \frac{Q}{u \sigma_z \sigma_y \pi} \, e^{-y^2/2\sigma_y^2} \, e^{-H_e^2/2\sigma_z^2}$$

Comparison of equations (14) and (24) yields this relationship between Sutton's dispersion coefficients and the dispersion coefficients in the generalized Gaussian distribution equation:

Gaussian	Sutton
σ_y^2	$\tfrac{1}{2} C_y^2 x^{2-n}$
σ_z^2	$\tfrac{1}{2} C_z^2 x^{2-n}$
$\sigma_y \sigma_z$	$\tfrac{1}{2} C_y C_z x^{2-n}$

In Sutton's equation (24), C_y and C_z are dispersion coefficients in units of (meters)$^{n/2}$ and n is a dimensionless stability parameter. In the Gaussian equation (14), σ_y and σ_z are dispersion coefficients in meters and their numerical values are functions of atmospheric stability and the downwind distance from the emission source.

THE PASQUILL DISPERSION COEFFICIENTS

In 1960, Meade[13] published some dispersion data estimates from which σ_y and σ_z could be derived for use in the Gaussian dispersion equations. Fairly soon thereafter, in 1961, Pasquill[8] published some similar estimates. Meade's publication indicated rather obliquely that his estimates were based on work done by Pasquill and others. Strangely enough, Pasquill's publication did not reference Meade's earlier publication. As shown in Table 3, a careful comparison of the two sets of published estimates reveals that **they are _not_ the same**, even though almost all subsequent publications by other authors seem to treat the two sets of estimates as being equivalent.

Both Meade and Pasquill presented their estimates in terms of (a) a plot of h versus the downwind distance x from the emission source, and (b) a tabulation of the lateral plume spread θ versus downwind distance x:

h = the vertical distance from the plume centerline up to either plume boundary point where the emissions concentration is 10 percent of the centerline concentration.

θ = the lateral plume spread (in the crosswind dimension) in degrees, corresponding to the two plume boundary points where the emissions concentrations amount to 10 percent of the centerline concentration.

Using the boundary-to-centerline concentration ratios tabulated on page 44 in Chapter 2, the above definitions of h and θ can be related to σ_z and σ_y as follows:

(25) $\quad h = 2.15 \, \sigma_z$

(26) $\quad \tan(\tfrac{1}{2}\theta) = (2.15 \, \sigma_y)/x$

Gifford[14] used equations (25) and (26) in a 1961 publication to develop plots of σ_y and σ_z versus downwind distance from the emission source for each of the six Pasquill stability classes A, B, C, D, E and F. In developing his plots, Gifford stated that he used Meade's estimates of h and θ. Those plots have become known as the "Pasquill-Gifford" dispersion coefficients.

Turner[9] later published his version of the Pasquill-Gifford dispersion coefficients. Since Turner's plots have gained very wide usage and acceptance, they have been reproduced herein as Figures 17 and 18.

Slade[7] and many others authors have published plots similar to Figures 17 and 18, and attributed them to Pasquill and Gifford. In fact, Slade provides two versions of the Pasquill-Gifford plots (see pages 102-103 and 408-409 in Slade's publication).

In 1974, Bowne[15] pointed out that there were some significant differences between Turner's plots and Slade's plots. To confirm Bowne's observation, Tables 4 and 5 herein were developed. A study of Tables 4 and 5 reveals that:

- Turner's plots of σ_y and σ_z represent the original Pasquill estimates[8] more faithfully than do any of the other plots.

- Slade's two sets of plots are not consistent with the original Pasquill estimates nor

TABLE 3

COMPARISON OF ORIGINAL PLUME SPREAD ESTIMATES

(by Pasquill[8] and Meade[13])

Stability Class	Vertical Spread			Horizontal Spread		
	Distance (km)	h, meters		Distance (km)	θ, degrees	
		Pasquill	Meade		Pasquill	Meade
A	0.1	30	30	0.1	60	50
A	0.5	220	240	1.0	47	42
A	1.0	1000	1300	10.0	33	34
A	1.7	3000	6500	100.0	20	25
B	0.1	23	23	0.1	45	40
B	1.0	235	260	1.0	37	34
B	5.0	1350	4800	10.0	28	27
B	10.0	2860	--	100.0	20	20
C	0.1	16	16	0.1	30	30
C	1.0	133	133	1.0	23	25
C	5.0	570	540	10.0	17	20
C	30.0	3000	2500	100.0	10	15
D				0.1	20	20
D				1.0	17	17
D				10.0	13	14
D				100.0	10	10

NOTE: This is a comparison of some selected values only, but it is sufficient to demonstrate the significant differences between Pasquill and Meade .... especially in the stability classes A, B and C.

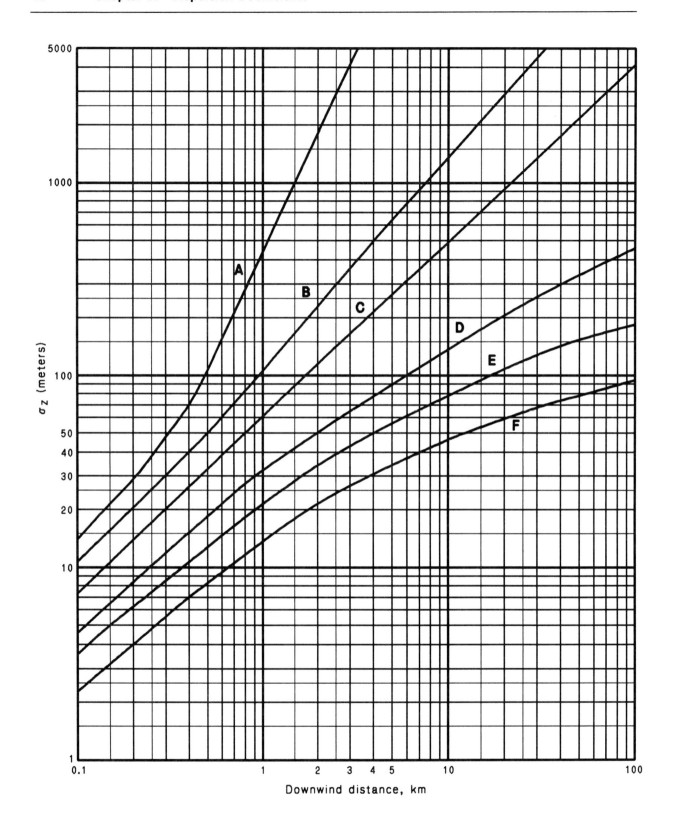

FIGURE 17

VERTICAL DISPERSION COEFFICIENTS

(Pasquill's rural values)

Chapter 3: Dispersion Coefficients 49

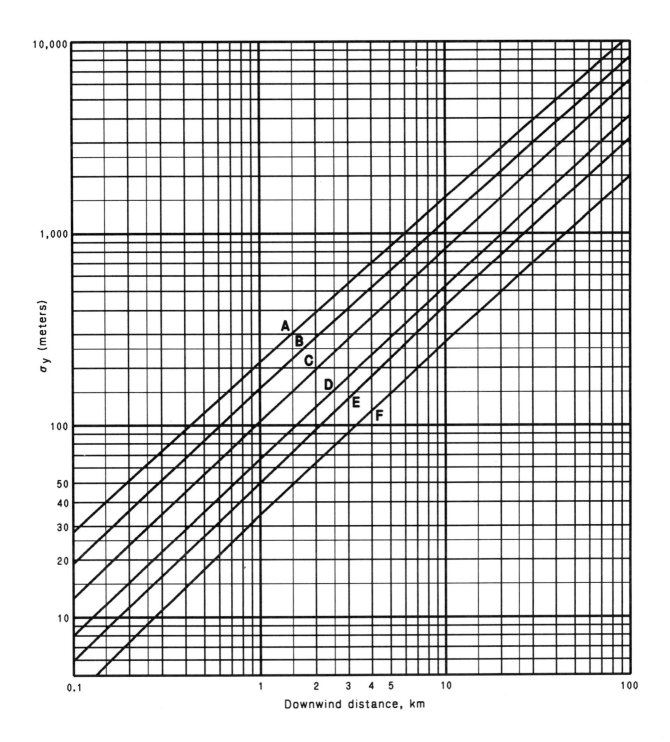

FIGURE 18

HORIZONTAL DISPERSION COEFFICIENTS
(Pasquill's rural values)

TABLE 4

COMPARISON OF LITERATURE σ_z VALUES WITH PASQUILL'S ORIGINAL VALUES

Distance (km)	Pasquill stability class[†]	Pasquill's original h (m)	Pasquill's original σ_z[††] (m)	Turner σ_z (m)	Slade (p. 103) σ_z (m)	Slade (p. 409) σ_z (m)	Gifford σ_z (m)	Bowne σ_z (m)
0.1	A	30	14	14	15	16	15	16
0.5	A	220	102	105	120	140	120	140
1.0	A	1000	465	450	540	750	540	650
1.7	A	3000	1395	1380	2900	3100	2900	2200
0.1	B	23	11	11	11	10	11	12
1.0	B	235	109	109	120	150	130	140
5.0	B	1350	628	630	1900	3000	2050	1900
10.0	B	2860	1330	1350	--	--	--	--
0.1	C	16	7.4	7.4	7.6	7.8	7.6	7.9
1.0	C	133	62	61	65	65	66	70
5.0	C	570	265	265	250	250	250	260
30.0	C	3000	1395	1350	930	1000	950	950
0.1	D	10	4.7	4.6	4.9	4.9	4.9	5.0
1.0	D	70	33	32	32	33	32	33
10.0	D	290	135	136	140	150	140	150
100.0	D	1000	465	460	430	450	450	490
0.1	E	7.5	3.5	3.5	3.5	3.0	3.5	3.8
1.0	E	46	21	22	23	24	22	24
10.0	E	170	79	78	84	83	89	85
100.0	E	390	181	183	180	180	185	190
0.1	F	5.0	2.3	2.3	2.3	1.5	2.3	2.4
1.0	F	30	14	14	14	14	14	15
10.0	F	100	47	46	47	48	48	49
100.0	F	200	93	94	90	94	92	91

[†] For stability class D, used Pasquill's curve labelled as D(2).
[††] Pasquill's h = 2.15 σ_z

TABLE 5

COMPARISON OF LITERATURE σ_y VALUES WITH PASQUILL'S ORIGINAL VALUES

Distance (km)	Pasquill stability class	Pasquill's original θ (deg)	Pasquill's original σ_y[†] (m)	Turner σ_y (m)	Slade (p. 102) σ_y (m)	Slade (p. 408) σ_y (m)	Gifford σ_y (m)	Bowne σ_y (m)
0.1	A	60.0	27	27	23	22	23	26
1.0	A	46.7	201	215	210	205	200	210
10.0	A	33.3	1391	1550	1500	1700	1600	1500
100.0	A	20.0	8201	11000	10000	12000	11000	12000
0.1	B	45.0	19	19	17	16	17	19
1.0	B	36.7	154	155	160	150	160	170
10.0	B	28.3	1173	1250	1200	1250	1200	1200
100.0	B	20.0	8201	8150	8200	9100	8500	9000
0.1	C	30.0	12.5	12.5	13.0	12.5	13.0	14.0
1.0	C	23.3	96	101	115	110	102	120
10.0	C	16.7	683	840	850	900	880	950
100.0	C	10.0	4069	6100	6100	6900	6400	6200
0.1	D	20.0	8.2	8.1	8.3	7.9	8.0	8.9
1.0	D	16.7	68	68	63	73	73	75
10.0	D	13.3	542	530	550	600	570	520
100.0	D	10.0	4069	4050	3800	4500	4200	4000
0.1	E	15.0	6.1	6.0	6.2	5.6	6.0	6.0
1.0	E	11.7	47	50	54	52	52	55
10.0	E	8.3	337	410	410	430	420	400
100.0	E	5.0	2031	3050	2750	3000	2850	2800
0.1	F	10.0	4.1	4.1	4.0	4.0	4.0	4.3
1.0	F	8.3	34	34	38	38	37	39
10.0	F	6.7	272	275	280	300	280	300
100.0	F	5.0	2031	2000	1900	2300	2050	2100

[†] $\tan(½\,\theta) = (2.15\,\sigma_y)/x$ where: x = downwind distance from source in meters

are they consistent with each other.

- Gifford's plots[14] are not consistent with the original Pasquill estimates.

- Bowne's plots[15] are not consistent with the original Pasquill estimates nor with either of Slade's plots. This is particularly surprising since Bowne reported the inconsistency between Turner's plots and Slade's plots, and also reported that Turner had faithfully represented Pasquill's estimates.

To ascertain how these various inconsistencies have developed, the original Pasquill estimates[8] are compared to Meade's estimates[13] in Table 3 herein. The comparison makes it obvious that Pasquill's estimates and Meade's estimates are not the same. The differences between Pasquill and Meade provide some reason for the many inconsistencies revealed in Tables 4 and 5, since Gifford used Meade's estimates and Turner used Pasquill's estimates. Some other factors which may have contributed to the inconsistencies are:

- Pasquill's θ values converged, at large downwind distances, for stability classes A and B, C and D, and E and F. Meade provided six sets of θ values at all downwind distances and hence no convergence was involved. To overcome Pasquill's convergent data, Turner seems to have used Pasquill's θ values for the B, D and F stability classes and to have adjusted Pasquill's θ values for the A, C and E classes so as to eliminate the convergence.

- The very small logarithmic plots published by Meade and by Gifford do not include the background grid lines and are very difficult to read accurately. This probably led to inconsistencies when subsequent authors attempted to reproduce Gifford's plots.

In summary, Turner's plots are faithful representations of the original Pasquill estimates and have gained wide usage. The Turner plots, reproduced as Figures 17 and 18 herein, will be used and referred to as the Pasquill dispersion coefficients in all the subsequent sections of this book.

In the book edited by Slade[7], an entire chapter is devoted to the many field experiments that have been performed to determine dispersion coefficients ... and is an excellent reference source on that subject. Slade's general conclusion is that the Pasquill dispersion coefficients are reasonably accurate for assessing the magnitude of dispersion problems.

Slade also repeats Pasquill's original caution that **the Pasquill dispersion coefficients are based on field experimental data that make them appropriate only for level terrain in open, rural areas.** In an urban or an industrial area, dynamic turbulence generated by buildings in the path of a dispersing plume tends to increase the spread of the plume. Buildings in the plume path also create so-called "downwash" effects which tend to bring the plume to ground level more quickly than would be the case in an open area with unobstructed flow. Urban and industrial areas are also generally somewhat warmer than rural areas and, therefore, have more thermally generated turbulence which further affects plume dispersion. For these reasons, the dispersion coefficients in Figures 17 and 18 are labelled as "Pasquill's rural values". Some suggested dispersion coefficients for use in urban areas are presented later in this chapter.

Chapter 3: Dispersion Coefficients

DISPERSION COEFFICIENTS IN EQUATION FORM

The use of computers for solving the Gaussian dispersion equations makes it very desirable to convert the Pasquill dispersion coefficient graphs in Figures 17 and 18 into analytical equations. Many different equations have been suggested for this purpose, one of which has been a power law function such as:

$$\sigma = ax^b$$

where x is the downwind distance from the emission source and the parameters a and b are functions of the atmospheric stability class and the downwind distance. Such power law functions have been suggested for both σ_z and σ_y.

Some of the U.S. EPA's computerized dispersion models have adopted a slightly different approach[16] by using a power law function for obtaining σ_z values and the following equation for obtaining σ_y values:

$$\sigma_y = (x/2.15) \tan \theta$$

where: $\theta = c - d(\ln x)$
ln = logarithm to the base e

and the parameters c and d are functions of the atmospheric stability class for any downwind distance x.

The most faithful representation, by far, of Turner's version of the rural Pasquill dispersion coefficients (Figures 17 and 18) is the equation published by McMullen[17]:

(27) $$\sigma = \exp[I + J(\ln x) + K(\ln x)^2]$$

where: σ = rural dispersion coefficient, m
x = downwind distance, km
exp[a] = e^a = 2.71828^a

The constants I, J and K provided by McMullen for use in equation (27) are:

TABLE 6

CONSTANTS I, J AND K FOR USE WITH EQUATION (27)

Pasquill Stability Class	For obtaining σ_z			For obtaining σ_y		
	I	J	K	I	J	K
A	6.035	2.1097	0.2770	5.357	0.8828	- 0.0076
B	4.694	1.0629	0.0136	5.058	0.9024	- 0.0096
C	4.110	0.9201	- 0.0020	4.651	0.9181	- 0.0076
D	3.414	0.7371	- 0.0316	4.230	0.9222	- 0.0087
E	3.057	0.6794	- 0.0450	3.922	0.9222	- 0.0064
F	2.621	0.6564	- 0.0540	3.533	0.9191	- 0.0070

A comparison is presented in Table 7 between the σ_y and σ_z values obtained from Turner's graphs (Figures 17 and 18) and those obtained from equation (27) as proposed by McMullen. It is obvious from a study of Table 7 that equation (27) provides σ_y and σ_z values which are essentially the same as those obtained from Turner's graphs.

TABLE 7

COMPARISON OF EQUATION (27) WITH FIGURES 17 AND 18

Pasquill Stability Class	Distance (km)	σ_z (meters)		Distance (km)	σ_y (meters)	
		Figure 17	Equation (27)		Figure 18	Equation (27)
A	0.1	14	14	0.1	27	27
A	0.5	105	111	1.0	215	212
A	1.0	430	418	10.0	1550	1555
A	1.7	1380	1383	100.0	11000	10523
B	0.1	11	11	0.1	19	19
B	1.0	109	109	1.0	155	157
B	5.0	630	626	10.0	1250	1194
B	10.0	1350	1358	100.0	8150	8185
C	0.1	7.4	7.3	0.1	12.5	12.1
C	1.0	61	61	1.0	101	105
C	5.0	265	267	10.0	840	833
C	30.0	1350	1362	100.0	6100	6111
D	0.1	4.6	4.7	0.1	8.1	7.9
D	1.0	32	30	1.0	68	69
D	10.0	136	140	10.0	530	549
D	100.0	460	463	100.0	4050	3993
E	0.1	3.5	3.5	0.1	6.0	5.8
E	1.0	22	21	1.0	50	51
E	10.0	78	80	10.0	410	408
E	100.0	183	187	100.0	3050	3081
F	0.1	2.3	2.3	0.1	4.1	4.0
F	1.0	14	14	1.0	34	34
F	10.0	46	47	10.0	275	273
F	100.0	94	90	100.0	2000	2023

NOTE: This is a comparison of some selected values only, but it is sufficient to show how very well equation (27) matches the Turner graphs of the Pasquill dispersion coefficients.

RURAL VERSUS URBAN DISPERSION COEFFICIENTS

Dispersing plumes encounter more turbulence in urban areas than in rural areas, due to the buildings as well as the somewhat warmer temperatures in urban areas. Higher turbulence also occurs in industrial plants densely populated with buildings or other structures. The additional turbulence created by an urban or industrial area is enough to alter the localized atmospheric stability to a less stable class than indicated by the prevailing meteorological conditions. In other words, if the prevailing meteorological conditions in an urban or industrial area indicate a class B stability, the increased turbulence would actually disperse a plume as if class A stability conditions prevailed. Thus, for any given set of meteorological conditions, the urban plume dispersion coefficients should be larger than the rural plume dispersion coefficients. Experimental data obtained by many investigators, notably McElroy and Pooler[18,19] and Shum, Loveland and Hewson[20] among others, have confirmed that urban areas have higher dispersion coefficients. As suggested by Bowne[15], the effect of urban areas upon plume dispersion could be dealt with by either: modifying the Pasquill stability classes to differentiate between urban and rural areas, or by using different dispersion coefficients for urban areas than for rural areas (without modifying the Pasquill stability classes). The latter option of using two sets of dispersion coefficients has been fairly widely adopted. In fact, some of the U.S. EPA's computerized dispersion models known as the RAM series include the option of using urban dispersion coefficients[21] derived from the experimental work of McElroy and Pooler.

Just what is the qualitative effect of higher urban dispersion coefficients upon plume dispersion? The higher coefficients cause an urban plume to spread more rapidly than a rural plume, and hence the maximum ground-level concentration of an urban plume occurs closer to the emission source than it does for a rural plume. If we used equation (15) to calculate plume ground-level concentrations versus downwind distance from the emission source:

$$(15) \quad C = \frac{Q}{u \sigma_z \sigma_y \pi} e^{-H_e^2 / 2 \sigma_z^2}$$

and if we kept Q, u, and H_e constant but used two sets of σ values (with one set lower than the other), a plot of the ground-level concentrations (C) versus downwind distance (x) would appear thus:

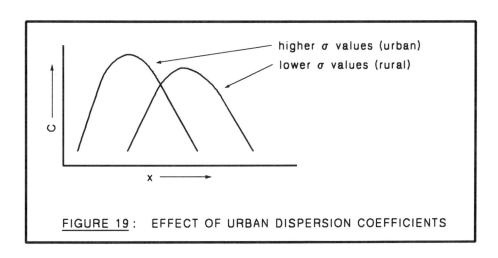

FIGURE 19: EFFECT OF URBAN DISPERSION COEFFICIENTS

As shown in Figure 19 on the prior page, the maximum ground-level concentration of an urban plume occurs closer to the emission source than is the case for a rural plume. As also shown in Figure 19, an urban plume's maximum ground-level concentration is generally slightly higher than the maximum ground-level concentration of a rural plume. Finally, at distances near to its emission source, an urban plume has higher ground-level concentrations than does a rural plume at the same distance from its source ... whereas, at distances far from the source, an urban plume has lower ground-level concentrations than does a rural plume at the same distance from its source.

In 1973, Briggs[22] developed what he called "plume half-widths" R_z and R_y from the previous experimental work of McElroy and Pooler in urban areas. Briggs defined his plume half widths in terms of the Gaussian dispersion coefficients as:

$$R = 1.25\,\sigma$$

In 1975, Gifford[23] restated Briggs' urban dispersion coefficients and developed the following equation for obtaining urban dispersion coefficients for use in Gaussian dispersion equations:

(28) $$\sigma = (Lx)(1 + Mx)^N$$

where: σ = urban dispersion coefficient, m
x = downwind distance, km

The constants L, M and N for use in equation (28) are:

TABLE 8

CONSTANTS L, M AND N FOR USE WITH EQUATION (28)

Pasquill Stability Class	For obtaining σ_z			For obtaining σ_y		
	L	M	N	L	M	N
A-B	240	1.00	0.50	320	0.40	-0.50
C	200	0.00	0.00	220	0.40	-0.50
D	140	0.30	-0.50	160	0.40	-0.50
E-F	80	1.50	-0.50	110	0.40	-0.50

The urban dispersion coefficients obtained from equation (28) are presented in Figures 20 and 21 to provide a direct comparison between Pasquill's rural dispersion coefficients and Briggs' urban dispersion coefficients as restated by Gifford.

Chapter 3: Dispersion Coefficients 57

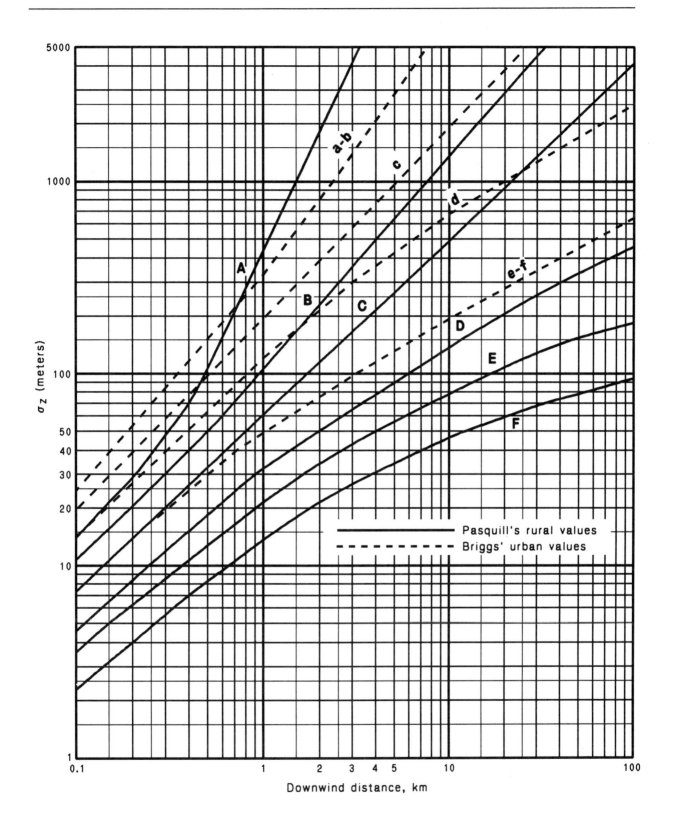

FIGURE 20

VERTICAL DISPERSION COEFFICIENTS
(Comparing rural and urban values)

Chapter 3: Dispersion Coefficients

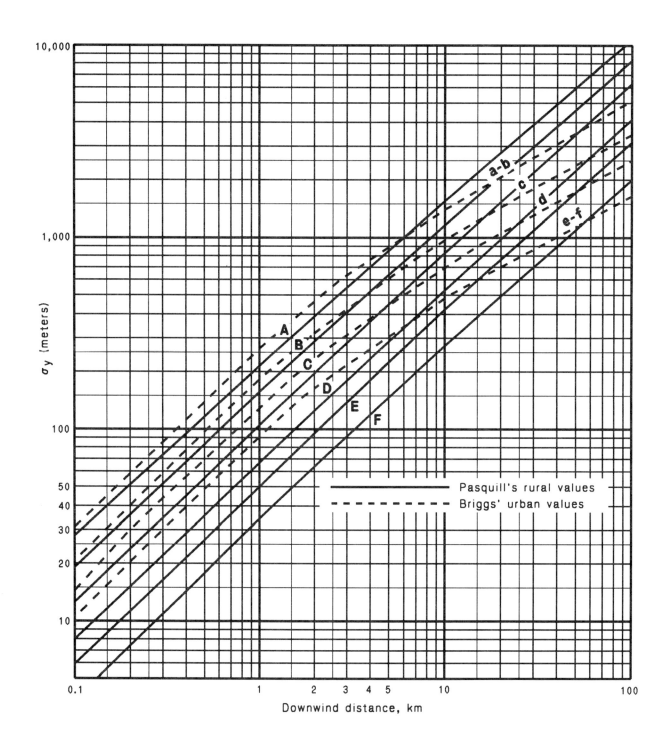

FIGURE 21

HORIZONTAL DISPERSION COEFFICIENTS
(Comparing rural and urban values)

Chapter 4

PLUME RISE

INTRODUCTION

A great many equations and mathematical models have been proposed for calculating the rise of stack gas plumes. Some of the proposed models are empirical while others have some theoretical basis. One of the earliest plume rise equations was developed by Bosanquet, Carey and Halton[24] in 1950 and was subsequently modified by Bosanquet.[25] Other plume rise equations and field data appeared in the early literature by: Holland (i.e., the Oak Ridge equation); Davidson (i.e., the Davidson-Byrant equation); Priestly and Ball; Schmidt; Sutton; Scorer; Thomas and co-workers at the TVA; Batchelor; and Morton, Taylor and Turner. More recent workers in the field of plume rise models include: Briggs; Stumke; Csanady; Moses; Strom; Lucas; Moore; Spurr; Carson; Rauch; Slawson; Smith; Singer; Brummage and many others.

In 1965, Briggs published his first plume rise model observations and comparisons.[26] In 1968, at a symposium sponsored by CONCAWE (a Dutch organization), he compared many of the models then available in the literature.[27] In that same year, Briggs also wrote the section of Slade's book[7] dealing with the comparative analyses of plume rise models. That was followed in 1969 by his classical critical review of the entire plume rise literature[28], in which he proposed a set of plume rise equations which have since become widely known as "the Briggs equations". Subsequently, Briggs modified his 1969 plume rise equations in 1971 and in 1972.[29,30]

The Bosanquet equations[25] gained wide acceptance in Canada and England and were used as the basis for regulatory guidelines and regulations involving industrial stack heights in those countries. The CONCAWE model[31] achieved some acceptance in Europe, particularly within the petroleum refining industry. In the United States, the following equations and models for the rise of buoyant stack gas plumes have all been used to some extent:

- The Holland equations[32]

- The ASME equations[33]

- The TVA equation[2]

- The Briggs equations[28,29,30]

All further discussion of plume rise models in this book is based on the Briggs equations. This does not constitute a value judgement that Briggs' equations are more valid than the other equations which have been proposed and used. It just reflects the fact that the U.S. EPA and many others have adopted Briggs' equations for use in their stack gas dispersion models. A 1976 survey[3] of 45 organizations (other than the EPA) engaged in stack gas dispersion modelling in the United States, Canada and Japan found that 43 percent of the organizations were then using the Briggs equations. If the federal EPA and local environmental agencies had been included, the percentage would probably have approached 70 percent. Since 1976, the Briggs equations have probably gained even wider usage.

PLUME RISE VARIABLES

At its point of origin from the exit of a stack or vent, a plume has vertical momentum[†] due to (a) its stack exit velocity and (b) its buoyancy if it is warmer than the surrounding ambient air. The magnitude of the exit gas momentum relative to the momentum of the ambient wind determines whether the plume will rise vertically or be bent over by the wind. Even if the plume is bent over, its trajectory (path) will continue to rise because the plume has a vertical velocity component retained from its initial stack exit velocity and buoyancy momentum.

By neglecting atmospheric friction (drag) and by assuming no heat loss (adiabatic conditions), the plume may be considered as retaining or "conserving" its original velocity and buoyancy momentum. However, the plume grows in mass by entraining ambient air. Hence, the plume's total velocity relative to the ambient wind velocity decreases even though its total momentum (i.e., mass times velocity) is conserved. The entrainment of air is caused by self-induced turbulence derived from the plume's initial velocity and buoyancy momentum, as well as by atmospheric turbulence. The further the plume travels, the more air it entrains and the slower its velocity becomes relative to the ambient wind velocity. Eventually, its vertical velocity component becomes essentially nil and the plume's trajectory levels off. Thus, the height of the plume trajectory at any given point is a function of the distance travelled from its origin to the given point.

As a broad generality, based on the above discussion of plume rise, we can list the major plume variables as being:

- The plume's initial velocity momentum at the stack exit
- The plume's initial buoyancy momentum at the stack exit
- Ambient wind velocity
- Downwind distance from plume source (for bent-over plumes)
- Atmospheric turbulence as categorized by the atmospheric stability class

TRAJECTORY OF A BENT-OVER BUOYANT PLUME

Briggs[28] divided plumes into these four general categories:

(1) Cold jet plumes in calm conditions
(2) Cold jet plumes in windy conditions
(3) Hot, buoyant plumes in calm conditions
(4) Hot, buoyant plumes in windy conditions

Briggs considered the trajectory of cold jet plumes to be dominated by their initial velocity momentum ... and the trajectory of hot, buoyant plumes to be dominated by their buoyancy momentum to the extent that their initial velocity momentum was relatively unimportant. Although Briggs proposed plume rise equations[28] for all four of the above plume categories, it is ***important to emphasize that "the Briggs equations" which have gained wide usage are those for bent-over, hot buoyant plumes.***

[†] Assuming the stack exit direction is vertical, as it is in most cases.

In general, Briggs' equations for bent-over, hot buoyant plumes are based on data and observations involving hot plumes from typical combustion sources such as the flue gas stacks from steam-generating boilers burning fossil fuels in large power plants. We can therefore assume stack exit velocities were in the range of 20-100 ft/sec (6-30 m/sec) with exit temperatures in the range of 250-500 °F (120-260 °C).

In his discussion of the trajectory of a bent-over, hot buoyant plume, Briggs[28] refers to an "initial stage", a "transition stage" and a "final stage". Unfortunately, his definition of those stages is not very clear. However, Figure 22 herein depicts the trajectory of Briggs' bent-over, hot buoyant plume as best as can be pieced together from his discussion. As shown in Figure 22:

- The total plume rise derives from its initial vertical velocity momentum and its initial buoyancy momentum.

- In the "initial stage" of plume rise, air entrainment results mostly from self-induced turbulence.

- In the "final stage" of plume rise, air entrainment results mostly from atmospheric turbulence since the plume's velocity relative to the ambient wind velocity has decreased to the point that the plume has very little self-induced turbulence.

- A "transitional stage" of plume rise is visualized between the initial and final stages.

- The downwind distance x^* is defined as the point at which the atmospheric turbulence begins to be the prime factor in the entrainment of air by the plume.

- At some multiple of x^*, taken as about 3.5 x^*, the maximum plume rise is assumed to have been attained.

THE BUOYANCY FLUX PARAMETER

Briggs uses a so-called buoyancy flux parameter, F, to categorize the stack exit buoyancy of a plume, and he defines that parameter as:

(a) $$F = \text{buoyancy force}/\pi \rho_a$$

The buoyancy force is readily derived from the difference in weight between a parcel of hot stack gas and the volume of ambient air which it displaces. Since the stack gas is warmer than the air:

(b) $$\begin{aligned}\text{buoyancy force} &= g(w_a - w_s) \\ &= g(\rho_a V_a - \rho_s V_s)\end{aligned}$$

where: g = gravitational acceleration of 9.807 m/sec^2
V = volume flow, m^3/sec
w = weight flow, g/sec
ρ = density, g/m^3
subscript s = stack gas
superscript a = displaced ambient air

Chapter 4: Plume Rise

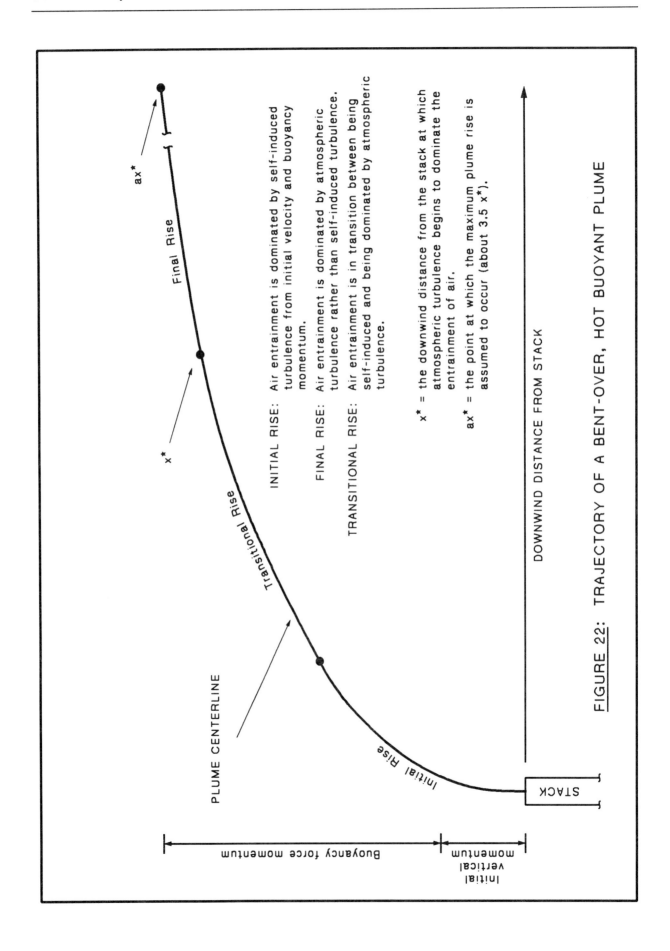

FIGURE 22: TRAJECTORY OF A BENT-OVER, HOT BUOYANT PLUME

and since V_a is the same as V_s:

(c) $\quad\quad$ buoyancy force $= gV_s(\rho_a - \rho_s)$

Combining equation (c) with Briggs' definition of F in equation (a):

(d) $\quad\quad F = (gV_s/\pi)(1 - \rho_s/\rho_a)$

Assuming that the stack gas molecular weight is essentially the same as that of air (which is typically so for a combustion flue gas):

(e) $\quad\quad \rho_a/\rho_s = T_s/T_a$

(f) $\quad\quad F = (gV_s/\pi)(1 - T_a/T_s)$
$\quad\quad\quad\quad = (gV_s/\pi)(T_s - T_a)/T_s$

and since:

(g) $\quad\quad V_s(T_s - T_a) = Q/(c_{ps}\rho_s)$

Equations (f) and (g) can be combined to obtain:

(h) $\quad\quad F = gQ/(\pi c_{ps} T_s \rho_s)$
$\quad\quad\quad\quad = gQ/(\pi c_{pa} T_a \rho_a)$

$$\begin{aligned}
\text{where:}\quad T &= \text{temperature, °K} \\
Q &= \text{stack gas sensible heat emission relative to} \\
  &\quad \text{the displaced air, in cal/sec} \\
c_{ps} &= \text{specific heat of stack gas, in cal/(g-°C)} \\
c_{pa} &= \text{specific heat of the ambient air, in cal/(g-°C)} \\
  &= c_{ps} \text{ (for typical combustion flue gases)} \\
T_s\rho_s &= T_a\rho_a \text{ per equation (e)}
\end{aligned}$$

The definition of Briggs' buoyancy flux parameter F can now be summarized in these three expressions, *which are completely equivalent to each other:*

(29a) $\quad\quad F = (gV_s/\pi)(T_s - T_a)/T_s$

(29b) $\quad\quad F = gv_s r^2 (T_s - T_a)/T_s$

(29c) $\quad\quad F = gQ/(\pi c_{pa} T_a \rho_a)$

$$\begin{aligned}
\text{where:}\quad g &= 9.807 \text{ m/sec}^2 \\
V_s &= \text{stack gas flow, m}^3/\text{sec} \\
v_s &= \text{stack exit velocity, m/sec} \\
r &= \text{stack exit radius, m} \\
Q &= \text{stack sensible heat emission, cal/sec} \\
c_{pa} &= \text{specific heat of ambient air, cal/(g-°C)} \\
\rho_a &= \text{ambient air density, g/m}^3 \\
T_a &= \text{ambient air temperature, °K} \\
T_s &= \text{stack gas temperature, °K} \\
F &= \text{buoyancy flux parameter, m}^4/\text{sec}^3
\end{aligned}$$

64   Chapter 4: Plume Rise

Equations (29 a, b and c) involve volumetric and heat flow rates (i.e., m^3/sec and cal/sec), which introduces the dimension of time. The word "flux" is a convention that is often used in naming parameters or variables that involve the dimension of time. In many cases, it denotes an energy flow rate through a unit surface (e.g., cal/sec/m^2).

Since g and π are constants, and since c_{pa}, T_a and ρ_a are essentially constants, it is obvious from equation (29c) that the buoyancy flux parameter is simply a measure of the stack gas sensible heat emission flow rate. Large combustion sources will result in large values of F, and small combustion sources will result in small values of F. For ambient air at 20 °C, equation (29c) provides these expressions of F in terms of the stack gas sensible heat emission Q :

(30a)   F in m^4/sec^3 = (3.68 × 10^{-5})(Q in cal/sec)

(30b)   F in m^4/sec^3 = (2.58 × 10^{-6})(Q in Btu/hr)

(30c)   F in m^4/sec^3 = (8.80)(Q in MW)

where: MW = megawatts

THE STABILITY PARAMETER

Briggs uses a so-called stability parameter, s, to categorize the effect of atmospheric turbulence on plume rise, and he defines that parameter as:

(31) $s = (g/T_a)d\theta/dz$

The stability parameter has also been defined as "the restoring force on a unit mass of air resulting from its vertical displacement from an equilibrium position".[7] To understand that definition, we must recall that $d\theta/dz$ is the potential temperature gradient ... which is the difference between the ambient temperature gradient and the adiabatic lapse rate , as defined by equation (4) in Chapter 1:

4) $d\theta/dz$ = ambient temperature gradient - adiabatic lapse rate
     = $dT/dz - \Gamma$

where Γ is the usual convention for the adiabatic lapse rate which is defined and discussed in Chapter 1.

Figure 23 depicts a parcel of air (the small circle enclosing the letter P) which is vertically displaced from altitude z_0 up to altitude z_1. Assuming that the air parcel expands and cools because of its displacement to a higher altitude and a resultant lower pressure, then its temperature T_p at z_1 is:

(a) $T_p = T_o + \Gamma \Delta z$

The temperature of the ambient air existing at altitude z_1 is:

(b) $T_a = T_o + (dT/dz)\Delta z$

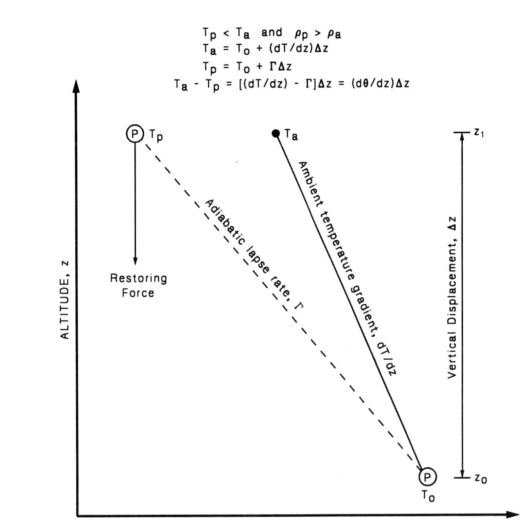

FIGURE 23: THE STABILITY PARAMETER

The difference in temperature between the risen air parcel and the existing ambient air at z_1 is therefore determined by the potential temperature gradient $d\theta/dz$:

(c) $$T_a - T_p = (dT/dz - \Gamma)\Delta z$$
$$= (d\theta/dz)\Delta z$$

The buoyancy force exerted on the risen air parcel at z_1 is the difference in weight between the risen air parcel and the volume of existing ambient air which it displaces:

(d) $$\text{Buoyancy force at } z_1 = ma$$
$$= gV(\rho_p - \rho_a)$$

and the equivalent buoyancy acceleration is obtained by dividing by the mass of the risen air parcel, $V\rho_p$, as shown in Figure 23:

(e) $$\text{Buoyancy acceleration, } a = g(1 - \rho_a/\rho_p)$$
$$= g(T_a - T_p)/T_a$$

Combining equations (c) and (e):

(f) $$\text{Buoyancy acceleration, } a = (g/T_a)(d\theta/dz)\Delta z$$

Finally, the restoring force on the risen air parcel per unit of mass per unit of vertical displacement is:

(g) $$\text{Restoring force, } s = ma$$
$$= (1)(g/T_a)(d\theta/dz)(1)$$

which is identical to the stability parameter defined by Briggs on the prior page.

Thus, we have derived Briggs' stability parameter:

(31) $$s = (g/T_a)d\theta/dz$$

where: s = stability parameter, sec^{-2}
g = 9.807 m/sec^2
$d\theta/dz$ = potential temperature gradient, °K/m
T_a = ambient air temperature, °K

For the example depicted in Figure 23, the potential temperature gradient is positive and the restoring force acts to return the risen air parcel downward. Hence, a positive potential temperature gradient acts to dampen turbulence and to create a stable atmosphere. A negative potential temperature gradient acts to enhance turbulence and to create an unstable atmosphere. A potential temperature gradient of zero acts neither to enhance or to dampen turbulence and thus acts to create a neutral atmosphere.

Therefore, it can be said that the stability parameter, s, is a measure of the effect of atmospheric turbulence upon the rise of a stack gas plume.

Table 9 below lists the defining ambient temperature gradients for each of the Pasquill stability classes as well as their equivalent potential temperature gradients:

TABLE 9: POTENTIAL TEMPERATURE GRADIENTS ($d\theta/dz$)

Stability Class[†]	Ambient Gradient	Adiabatic lapse rate	Potential gradient	
	dT/dz	Γ	$d\theta/dz = dT/dz - \Gamma$	
	(°F/1000 ft)	(°F/1000 ft)	(°F/1000 ft)	(°K/m)
A	<-10.4	-5.5	<-4.9	<-0.009
B	-9.9	-5.5	-4.4	-0.008
C	-8.8	-5.5	-3.3	-0.006
D	-5.5	-5.5	0.0	0.000
E	2.8	-5.5	8.3	0.015
F	>8.2	-5.5	>13.7	>0.025

[†] A is the most unstable class and F is the most stable class.

THE MOMENTUM FLUX PARAMETER

In his plume rise equations for cold jet plumes, Briggs uses a so-called momentum flux parameter, F_m, which he defines as the stack exit gas velocity momentum divided by $\pi\rho_a$.

The velocity momentum (i.e., mass × velocity) of the stack exit gas is:

(a) $\quad$ Momentum $= mv$

$$= (\rho_s V_s)(v_s) = \rho_s(\pi r^2 v_s)(v_s)$$

Therefore, Briggs' momentum flux parameter is:

(b) $\quad F_m = \text{momentum}/\pi\rho_a$

$$= (\rho_s/\rho_a) r^2 v_s^2$$

Assuming that the stack gas molecular weight is essentially the same as that of air, which is typically so for a combustion flue gas:

(32) $\quad F_m = (T_a/T_s) r^2 v_s^2$

where: T_a = ambient air temperature, °K
T_s = stack gas temperature, °K
V_s = stack gas flow, m³/sec
v_s = stack exit velocity, m/sec
r = stack exit radius, m
F_m = momentum flux parameter, m⁴/sec²

BRIGGS' EQUATIONS FOR BENT-OVER BUOYANT PLUMES

Hopefully, the reader has now gained an insight into the many variables involved in predicting the rise of stack gas plumes, and has a good understanding of the derivation and physical meaning of the buoyancy, stability and momentum parameters. It should also be clear that the mathematical expression of those parameters, in equations (29), (31) and (32), involves ***the explicit assumption that the plume consists of typical fossil fuel combustion products with an average molecular weight and specific heat that are essentially the same as for air.***

Based upon a dimensional analysis of the variables involved as well as a theoretical analysis, Briggs developed a set of plume rise equations for bent-over, hot buoyant plumes which he then correlated and compared with various field observations and data.[28] Subsequently, he modified those equations in 1971 and again in 1972.[29,30] Unfortunately, literature references to Briggs' equations by other authors rarely make clear which version of his equations are referenced. In those cases where the specific equations are given by other authors, Briggs' modifications of 1971 and 1972 are often found not to have been incorporated.

> *For example, a publication by Moses and Kraimer[34] in late 1972 was quite clearly referencing Briggs' 1969 equations despite the fact that Briggs had modified them quite significantly in 1971 and had again made a minor change in early 1972.*
>
> *As another example, a 1977 paper[35] includes one of Briggs' equations for plume rise in a stable atmosphere. However, the equation uses a numerical coefficient from Briggs' 1969 version rather than his later 1972 version. The paper also fails to qualify the given equation as being applicable for the maximum plume rise rather than for the transitional stage of rise.*
>
> *The same lack of clarity occurs in another paper[36] which gives two of Briggs' equations for plume rise in neutral and stable atmospheres without noting that the equations are for the maximum plume rise rather than for the transitional stage of rise.*

In May of 1973, the EPA revised their version of the Briggs equations to reflect his 1971 and 1972 modifications. The EPA's memorandum of revision[37] included a computer program subroutine which was to be used in modifying all of the EPA's pertinent dispersion models "in the near future".

The definition and evolution of Briggs' plume rise equations for bent-over, buoyant plumes are best discussed in the chronological order of his 1969, 1971 and 1972 publications.

1969 -- Briggs proposed this set of plume rise equations for bent-over, buoyant plumes:[28]

Unstable or neutral atmospheric conditions:

- Stack sensible heat emission of 20 MW or more:

$$\Delta h = 1.6 \, F^{1/3} \, x^{2/3} \, u^{-1} \qquad \text{for } x < 10 \, h_s$$

$$\Delta h_{max} = 1.6 \, F^{1/3} \, (10 \, h_s)^{2/3} \, u^{-1} \qquad \text{for } x > 10 \, h_s$$

- Stack sensible heat emission of less than 20 MW:

$$\Delta h = 1.6\, F^{1/3}\, x^{2/3}\, u^{-1} \qquad \text{for } x < 3\, x^*$$

$$\Delta h_{max} = 1.6\, F^{1/3}\, (3\, x^*)^{2/3}\, u^{-1} \qquad \text{for } x > 3\, x^*$$

$$\text{where:}\quad 3\, x^* = 6.5\, F^{0.4}\, h_s^{0.6} \qquad \text{for } h_s < 305 \text{ meters}$$

$$3\, x^* = 202\, F^{0.4} \qquad \text{for } h_s > 305 \text{ meters}$$

Stable atmospheric conditions:

$$\Delta h = 1.6\, F^{1/3}\, x^{2/3}\, u^{-1} \qquad \text{for } x < 2.4\, u\, s^{-1/2}$$

$$\Delta h_{max} = 2.9\, (F/us)^{1/3} \qquad \text{for } x > 2.4\, u\, s^{-1/2}$$

where: Δh = plume rise for first and transitional stages, m
Δh_{max} = maximum plume rise, m
h_s = actual stack height, m
x = downwind distance from stack, m
u = wind velocity, m/sec
F = buoyancy parameter, m^4/sec^3 per equation (29) or (30)
s = stability parameter, sec^{-2} per equation (31)

As can be seen in the above set of six equations, Briggs' 1969 plume rise equation for the first and transitional stages of rise (see Figure 22) was the same for all atmospheric conditions (unstable, neutral or stable), namely:

$$\Delta h = 1.6\, F^{1/3}\, x^{2/3}\, u^{-1}$$

However, Briggs proposed three different equations for the maximum plume rise, as well as three different definitions of the downwind distance from the stack to the point at which the maximum plume rise occurs:

° For "large" heat emission stacks in unstable or neutral atmospheric conditions, the maximum plume rise was defined as occurring at $x = 10\, h_s$

° For "small" heat emission stacks in unstable or neutral atmospheric conditions, the maximum plume rise was defined as occurring at $x = 3.0\, x^*$, and x^* was defined as a function of the buoyancy parameter F and of the actual stack height h_s

° For a stack heat emission of any size in stable atmospheric conditions, the maximum plume rise was defined as occurring at the point where $x = 2.4\, u\, s^{-1/2}$

Briggs' 1969 definition of a large heat emission stack was 20 MW or more of stack gas sensible heat emission. That equates to:

° A buoyancy flux parameter of 176 m^4/sec^3 or more
° 68×10^6 Btu/hr or more of stack gas sensible heat
° 4.8×10^6 cal/sec or more of stack gas sensible heat

1971 -- Briggs proposed this revised set of plume rise equations for bent-over, buoyant plumes:[29]

Unstable or neutral atmospheric conditions:

- Buoyancy parameter $F > 55$ m^4/sec^3 :

    $\Delta h = 1.6\ F^{1/3}\ x^{2/3}\ u^{-1}$ for $x < 3.5\ x^*$

    $\Delta h_{max} = 1.6\ F^{1/3}\ (3.5\ x^*)^{2/3}\ u^{-1}$ for $x > 3.5\ x^*$

    where: $3.5\ x^* = 119\ F^{0.4}$

- Buoyancy parameter < 55 m^4/sec^3 :

    $\Delta h = 1.6\ F^{1/3}\ x^{2/3}\ u^{-1}$ for $x < 3.5\ x^*$

    $\Delta h_{max} = 1.6\ F^{1/3}\ (3.5\ x^*)^{2/3}\ u^{-1}$ for $x > 3.5\ x^*$

    where: $3.5\ x^* = 49\ F^{0.625}$

Stable atmospheric conditions:

$\Delta h = 1.6\ F^{1/3}\ x^{2/3}\ u^{-1}$ for $x < 2\ u\ s^{-1/2}$

$\Delta h_{max} = 2.9\ (F/us)^{1/3}$ for $x > 2\ u\ s^{-1/2}$

As can be seen in the above set of six equations, Briggs again proposed the "2/3 law" plume rise equation just as he had done in 1969 for the first and transitional stages of plume rise during all atmospheric conditions (unstable, neutral or stable), namely:

$\Delta h = 1.6\ F^{1/3}\ x^{2/3}\ u^{-1}$

However, Briggs changed his three 1969 equations for the maximum plume rise, as well as changing his definitions of the downwind distance from the stack to the point at which the maximum plume rise occurs:

º For "large" heat emission stacks (i.e., $F > 55$ m^4/sec^3) in unstable or neutral atmospheric conditions, the point of maximum plume rise was defined as occurring at $x = 3.5\ x^* = 119\ F^{0.4}$

º For "small" heat emission stacks (i.e., $F < 55$ m^4/sec^3) in unstable or neutral atmospheric conditions, the point of maximum plume rise was defined as occurring at $x = 3.5\ x^* = 49\ F^{0.625}$

º For a stack heat emission of any size in stable atmospheric conditions, the maximum plume rise was defined as occurring at $x = 2\ u\ s^{-1/2}$

Briggs' 1971 definition of a large heat emission stack was considerably smaller than it had been in 1969, and it equates to these four equivalent values:

- A buoyancy flux parameter F of 55 m^4/sec^3 or more
- 6.25 MW or more of stack gas sensible heat
- 21×10^6 Btu/hr or more of stack gas sensible heat
- 1.5×10^6 cal/sec or more of stack gas sensible heat

1972 -- Briggs made a minor revision to his 1971 plume rise equations for bent-over, buoyant plumes.[30]

Briggs proposed that the numerical coefficient in his maximum plume rise equation for stable conditions be changed from 2.9 to 2.4, so that the 1972 version became:

$$\Delta h_{max} = 2.4 \, (F/us)^{1/3} \qquad \text{for } x > 2 \, u \, s^{-1/2}$$

The above chronological presentation briefly summarizes Briggs' plume rise publications up to about 1980. In some of the above presentation, Briggs' original equations were converted to metric units so that all of his various equations could be compared on a consistent basis.

As an extension of his equations, **Briggs suggested[29] that his 1971 equations for neutral and unstable atmospheric conditions could also be used for stable conditions if they result in a lower plume rise than his equation for maximum plume rise in stable conditions.** That statement can be developed mathematically by equating the three 1971 equations for Δh_{max}:

$$1.6 \, F^{1/3} \, x^{2/3} \, u^{-1} = 2.4 \, (F/us)^{1/3} = 1.6 \, F^{1/3} \, (3.5 \, x^*)^{2/3} \, u^{-1}$$

We find that $x = 1.84 \, u \, s^{-1/2}$ when the first two equations are equal. And we also find that $3.5 \, x^* = 1.84 \, u \, s^{-1/2}$ when the last two equations are equal. Thus, whenever

$$3.5 \, x^* > 1.84 \, u \, s^{-1/2} < x$$

then $2.4 \, (F/us)^{1/3}$ will give the lowest value of Δh_{max} and will apply for stable atmospheric conditions. Therefore, these equations apply for use in stable atmospheric conditions:

- When $1.84 \, u \, s^{-1/2} > 3.5 \, x^*$:

    $\Delta h = 1.6 \, F^{1/3} \, x^{2/3} \, u^{-1}$ \qquad for $x < 3.5 \, x^*$

    $\Delta h_{max} = 1.6 \, F^{1/3} \, (3.5 \, x^*)^{2/3} \, u^{-1}$ \qquad for $x > 3.5 \, x^*$

- When $1.84 \, u \, s^{-1/2} < 3.5 \, x^*$:

    $\Delta h = 1.6 \, F^{1/3} \, x^{2/3} \, u^{-1}$ \qquad for $x < 1.84 \, u \, s^{-1/2}$

    $\Delta h_{max} = 2.4 \, (F/us)^{1/3}$ \qquad for $x > 1.84 \, u \, s^{-1/2}$

    where: $3.5 \, x^* = 119 \, F^{0.4}$ \qquad for $F > 55 \, m^4/sec^3$

    $3.5 \, x^* = 49 \, F^{0.625}$ \qquad for $F < 55 \, m^4/sec^3$

Before recapitulating all of Briggs' equations in their 1972 version, we will examine the magnitude of combustion sources consistent with a buoyancy parameter F of 55 m^4/sec^3. Since the buoyancy parameter is a function of the stack gas <u>sensible</u> heat emission, it is therefore also a function of the combustion unit thermal efficiency and the fuel composition. We will start with an F of 55 m^4/sec^3 being equivalent to a stack gas sensible heat emission of 6.25 megawatts per equation (30c). If we assume the following:

- A fuel net heating value that is 94 percent of its gross heating value
- A combustion efficiency of 88 percent based upon the fuel's gross heating value
- A power plant fired with a fossil fuel and having an overall efficiency of 34 percent based upon the fuel's gross heating value

Figure 24 shows that:

$$\text{An F of 55 } m^4/sec^3 = 6.25 \text{ MW of stack gas sensible heat emission}$$
$$= 356,000,000 \text{ Btu/hr of fuel burned}$$
$$= 35 \text{ MW of generated electrical energy}$$

As a generalized approximation, the stack gas sensible heat emission from a power plant firing a fossil fuel is 15-20 percent of the generated output of electrical energy. In terms of today's modern power plants of 1000 MW output or more, an F of 55 m^4/sec^3 is equivalent to quite a small power plant.

Recapitulation -- The Briggs equations for bent-over, buoyant plumes are recapitulated and summarized below in their 1972 version:

PASQUILL STABILITY CLASSES A, B, C, D (unstable and neutral atmospheric conditions)

When $F \geq 55$ m^4/sec^3 :

(33) $\quad \Delta h = 1.6 \, F^{1/3} \, x^{2/3} \, u^{-1}$ $\qquad\qquad$ for $x < x_f$

(34) $\quad \Delta h_{max} = 1.6 \, F^{1/3} \, x_f^{2/3} \, u^{-1} = 38.7 \, F^{0.60} \, u^{-1}$ $\qquad\qquad$ for $x \geq x_f$

When $F < 55$ m^4/sec^3 :

(35) $\quad \Delta h = 1.6 \, F^{1/3} \, x^{2/3} \, u^{-1}$ $\qquad\qquad$ for $x < x_f$

(36) $\quad \Delta h_{max} = 1.6 \, F^{1/3} \, x_f^{2/3} \, u^{-1} = 21.4 \, F^{0.75} \, u^{-1}$ $\qquad\qquad$ for $x \geq x_f$

PASQUILL STABILITY CLASSES E, F (stable atmospheric conditions)

When $1.84 \, u \, s^{-1/2} \geq x_f$:

(37) $\quad \Delta h = 1.6 \, F^{1/3} \, x^{2/3} \, u^{-1}$ $\qquad\qquad$ for $x < x_f$

(38) $\quad \Delta h_{max} = 1.6 \, F^{1/3} \, x_f^{2/3} \, u^{-1} = 38.7 \, F^{0.60} \, u^{-1}$ $\qquad\qquad$ for $x \geq x_f$ and $F \geq 55$ m^4/sec^3

(39) $\quad \Delta h_{max} = 1.6 \, F^{1/3} \, x_f^{2/3} \, u^{-1} = 21.4 \, F^{0.75} \, u^{-1}$ $\qquad\qquad$ for $x \geq x_f$ and $F < 55$ m^4/sec^3

When $1.84\, u\, s^{-1/2} < x_f$:

(40) $\quad \Delta h = 1.6\, F^{1/3}\, x^{2/3}\, u^{-1}$ $\qquad$ for $x < 1.84\, u\, s^{-1/2}$

(41) $\quad \Delta h_{max} = 2.4\, (F/us)^{1/3}$ $\qquad$ for $x \geq 1.84\, u\, s^{-1/2}$

where:
- Δh = plume rise for first and transitional stages, m
- Δh_{max} = maximum plume rise, m
- x = downwind distance from stack, m
- x_f = downwind distance from stack to maximum (final) plume rise, m
- x^* = downwind distance to end of transitional rise stage, m
- $x_f = 3.5\, x^* = 119\, F^{0.40}$ for $F \geq 55\, m^4/sec^3$
- $\qquad\quad\ = 49\, F^{0.625}$ for $F < 55\, m^4/sec^3$
- u = wind velocity, m/sec
- F = buoyancy parameter, m^4/sec^3 per equation (29) or (30)
- s = stability parameter, sec^{-2} per equation (31)

CALCULATED PLUME TRAJECTORIES

The use of Briggs' equations for bent-over, buoyant plumes is illustrated with some typical plume rise calculations in Examples 2, 3 and 4. The examples were selected so as to:

- Use most of the Briggs equations (33) through (41) as summarized just above.
- Illustrate the effect of wind velocity on plume rise.
- Emphasize the effect of atmospheric stability on the maximum plume rise, Δh_{max}.

Example 2 is based on a stack gas emission from a process furnace in a petroleum refinery. The buoyancy flux parameter F of 40 m^4/sec^3 is equivalent to a sensible heat emission of 15×10^6 Btu/hr as per equation (30b). Assuming the furnace has a thermal efficiency of 85 percent, that sensible heat emission corresponds to burning 207×10^6 Btu/hr of fuel in the furnace. Example 2 illustrates the calculation of an effective stack height (H_e) for a "small" sensible heat emission in an unstable atmosphere at a downwind distance from the stack larger than x_f, so that equation (36) for the maximum plume rise is applicable.

Example 3 is based upon a stack gas emission from a large steam-generating boiler, with an F of 611 m^4/sec^3 which is equivalent to a sensible heat emission of 237×10^6 Btu/hr. Part (A) of example 3 uses equations (40) and (41) to calculate the plume rise trajectory for that large heat emission in a stable atmosphere at various downwind distances from the source stack. Part (B) of example 3 repeats the same calculations at a lower wind velocity. Example 4 repeats the trajectory calculations for the same plume in an unstable atmosphere.

The calculated results of examples 3 and 4 are plotted in Figure 25, which shows that:

- Lower windspeeds result in higher plume rises, since Briggs' Δh is inversely proportional to windspeed during the first stage and transitional stage of rise.

- For a given stack emission at a given windspeed, the maximum plume rise is lower in a stable atmosphere than in an unstable atmosphere. The maximum plume rise also occurs closer to the source stack in a stable atmosphere than in an unstable one.

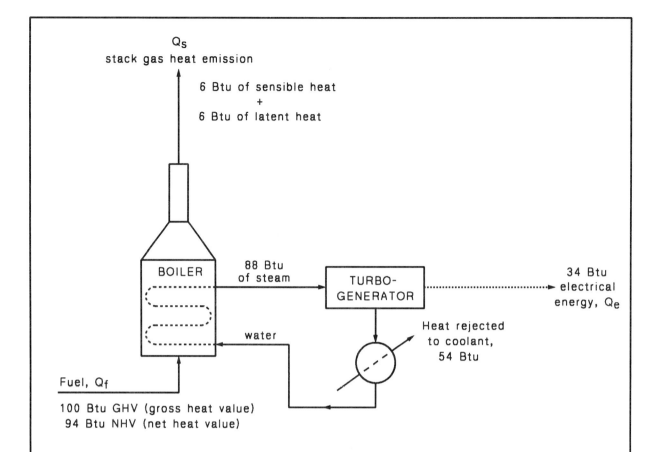

Since a buoyancy flux parameter of $F = 55$ m^4/sec^3 is equivalent to a stack gas sensible heat emission of 6.25 megawatts:

	Q_f	GHV Efficiency, %		Stack Gas Q_s		Q_e
		Boiler	Overall	Sensible	Latent	
Megawatts ---	104.17	88	34	6.25	6.25	35.41
10^6 Btu/hr ---	355.74	88	34	21.34	21.34	120.94

Thus, an F of 55 m^4/sec^3 is equivalent to the stack gas sensible heat emission from a power plant producing about 35 MW of electrical energy from the combustion of about 356,000,000 Btu/hr of fuel.

FIGURE 24

STACK GAS HEAT EMISSION
vs
COMBUSTION SOURCE MAGNITUDE

EXAMPLE 2: CALCULATION OF EFFECTIVE STACK HEIGHT
(for a bent-over, hot buoyant plume)

Calculate the effective stack height, H_e, of a stack gas plume centerline at a downwind distance of x = 2500 ft (762 m) from the exit of the source stack.

GIVEN: Stack exit gas flow = 400 SCF/sec[†]
Stack exit temperature = 650 °F   = 616 °K
Stack height, h_s = 250 ft   = 76 m
Ambient temperature = 60 °F   = 289 °K
Wind velocity, u = 2 ft/sec = 0.61 m/sec
Atmospheric stability = Pasquill class B

Convert the stack exit gas flow into m^3/sec at the stack exit temperature:

V_s = 400(616/289)/(35.31 ft^3/m^3)
     = 24.15 m^3/sec at 616 °K and 1 atmosphere pressure

Calculate the buoyancy factor, F, using equation using equation (29a):

F = (9.807)(24.15/π)(616 - 289)/616
  = 40.02 m^4/sec^3

Since F < 55 m^4/sec^3:

x_f = 49 $F^{0.625}$ (see page 73)
     = 49(40.02)$^{0.625}$
     = 492 m

Since x = 762 m, then x > x_f.

Since x > x_f, F < 55 m^4/sec^3, and the stability class is B, then equation (36) is applicable for calculating Δh:

Δh = 21.4(40.02)$^{0.75}$(0.61)$^{-1}$
         = 558 m

H_e = h_s + Δh (see page 38)
     = actual stack height + plume rise at 762 m downwind from source stack
     = 76 + 558
     = 634 m (or 2,080 ft)

[†] Standard cubic feet (SCF) measured at 60 °F (289 °K) and 1 atmosphere (14.696 psia)

76 Chapter 4: Plume Rise

EXAMPLE 3: CALCULATION OF COMPLETE PLUME TRAJECTORIES (for bent-over, hot buoyant plumes)

(A) Calculate the plume rise, Δh, at various downwind distances from the source stack to establish the complete trajectory of a stack gas plume from a boiler plant.

GIVEN:
Stack gas flow = 5.2×10^6 pounds/hr
Stack gas specific heat = 0.24 Btu/(lb-°F)
Stack exit temperature = 240 °F = 389 °K
Ambient temperature = 50 °F = 283 °K
Wind velocity, u = 13 ft/sec = 4 m/sec
Ambient temperature gradient = dT/dz = 3 °F/1000 ft

Calculate the stack gas sensible heat emission relative to ambient air:

$Q = (5.2 \times 10^6 \text{ lbs/hr})(0.24)(240 - 50) = 237 \times 10^6$ Btu/hr of sensible heat

Calculate the buoyancy factor, F, using equation (30b):

$F = (2.58 \times 10^{-6})(237 \times 10^6 \text{ Btu/hr}) = 611 \text{ m}^4/\text{sec}^3$

Determine the Pasquill stability class:

dT/dz = 3 °F/1000 ft
= Pasquill stability class E per Table 1, page 8

Calculate the potential temperature gradient, $d\theta/dz$:

$d\theta/dz = dT/dz - \Gamma$ (see page 64)
$= 3 - (-5.5) = 8.5$ °F/1000 ft
$= 8.5(3.281 \text{ ft/m})/(1.8 \text{ °F/°K})(1000) = 0.0155$ °K/m

Calculate stability parameter, s, from equation (31):

$s = (9.807/283)(0.0155) = 0.000537 \text{ sec}^{-2}$

Calculate $1.84 \, u \, s^{-1/2}$:

$= 1.84(4)(0.000537)^{-1/2} = 318$ m

Since $F > 55 \text{ m}^4/\text{sec}^3$:

$x_f = 119 \, F^{0.40}$ (see page 73)
$= 119(611)^{0.40} = 1{,}549$ m

Summary of parameters to this point:

$F = 611 \text{ m}^4/\text{sec}^3$ $1.84 \, u \, s^{-1/2} = 318$ m
$u = 4$ m/sec $s = 0.000537 \text{ sec}^{-2}$
$x_f = 1{,}549$ m   stability = class E

Since the atmospheric stability is class E and $1.84 \, u \, s^{-1/2} < x_f$, then either equation (40) or equation (41) applies to the calculation of the plume rise, Δh. First, calculate Δh at downwind distances of 25, 50, 100 and 200 m using equation (40) since $x < 1.84 \, u \, s^{-1/2}$ at those distances:

$\Delta h = 1.6 \, (611)^{1/3}(25)^{2/3}(4)^{-1} = 29$ m

$\Delta h = 1.6 \, (611)^{1/3}(50)^{2/3}(4)^{-1} = 46$ m

$\Delta h = 1.6 \, (611)^{1/3}(100)^{2/3}(4)^{-1} = 73$ m

$\Delta h = 1.6 \, (611)^{1/3}(200)^{2/3}(4)^{-1} = 116$ m

Chapter 4: Plume Rise

EXAMPLE 3 cont'd:

Next, calculate Δh_{max} at downwind distance of 318 m using equation (41) since $x \geq 1.84\, u\, s^{-1/2}$ at that distance:

$\Delta h_{max} = 2.4[611(1/4)(1/0.000537)]^{1/3} = 158$ m

Summary of part (A) results:

	x meters downwind				
	25	50	100	200	318
Δh, meters	29	46	73	116	158

(B) Repeat the trajectory calculation in part (A) using a wind velocity of 6.5 ft/sec which is equivalent to 2 m/sec:

$1.84\, u\, s^{-1/2} = 1.84(2)(0.000537)^{-1/2} = 159$ m

All of the other parameters are not affected by a change in the wind velocity and, thus, remain the same as in part (A):

$F = 611$ m^4/sec^3 $1.84\, u\, s^{-1/2} = 159$ m
$u = 2$ m/sec $s = 0.000537$ sec^{-2}
$x_f = 1,549$ m   stability = class E

Since the atmospheric stability is class E and $1.84\, u\, s^{-1/2} < x_f$, then either equation (40) or equation (41) applies to the calculation of the plume rise, Δh. Calculations similar to those in part (A) yield this summary:

	x meters downwind			
	25	50	100	159
Δh, meters	58	92	146	199

A plot of the trajectories obtained in parts (A) and (B) is presented in Figure 25.

78   Chapter 4: Plume Rise

EXAMPLE 4: CALCULATION OF COMPLETE PLUME TRAJECTORIES
(for bent-over, hot buoyant plumes)

(A) Calculate the plume rise trajectory of the same boiler plant plume as in Example 3, except use an ambient temperature gradient (dT/dz) of -10 °F/1000 ft.

GIVEN:
Stack gas flow = 5.2×10^6 pounds/hr
Stack gas specific heat = 0.24 Btu/(lb-°F)
Stack exit temperature = 240 °F = 389 °K
Ambient temperature = 50 °F = 283 °K
Wind velocity, u = 13 ft/sec = 4 m/sec
Ambient temperature gradient = dT/dz = -10 °F/1000 ft

Calculate the stack gas sensible heat emission relative to ambient air:

$Q = (5.2 \times 10^6 \text{ lbs/hr})(0.24)(240 - 50) = 237 \times 10^6$ Btu/hr of sensible heat

Calculate the buoyancy factor, F, using equation (30b):

$F = (2.58 \times 10^{-6})(237 \times 10^6 \text{ Btu/hr}) = 611 \text{ m}^4/\text{sec}^3$

Determine the Pasquill stability class:

dT/dz = -10 °F/1000 ft
= Pasquill stability class B per Table 1, page 8

[Since the stability class is B, the potential temperature gradient dθ/dz and the stability parameter s are not required to calculate the plume rise trajectory]

Since F > 55 m^4/sec^3:

$x_f = 119 \, F^{0.40}$ (see page 73)
$\quad = 119(611)^{0.40} \qquad\qquad = 1,549$ m

Summary of parameters to this point:

F = 611 m^4/sec^3 stability = class B
u = 4 m/sec x_f = 1,549 m

Since the atmospheric stability is class B and F > 55 m^4/sec^3, then either equation (33) or equation (34) applies to the calculation of the plume rise, Δh. First, calculate Δh at downwind distances of 25, 50, 100, 200, 600, and 1200 m using equation (33) since x < x_f at those distances:

$\Delta h = 1.6 \, (611)^{1/3}(25)^{2/3}(4)^{-1} = 29$ m

$\Delta h = 1.6 \, (611)^{1/3}(50)^{2/3}(4)^{-1} = 46$ m

$\Delta h = 1.6 \, (611)^{1/3}(100)^{2/3}(4)^{-1} = 73$ m

$\Delta h = 1.6 \, (611)^{1/3}(200)^{2/3}(4)^{-1} = 116$ m

$\Delta h = 1.6 \, (611)^{1/3}(600)^{2/3}(4)^{-1} = 241$ m

$\Delta h = 1.6 \, (611)^{1/3}(1200)^{2/3}(4)^{-1} = 383$ m

Next, calculate Δh_{max} at downwind distance of 1,549 m using equation (34) since x ≥ x_f at that distance:

$\Delta h_{max} = 38.7(611)^{0.60}(4)^{-1} = 454$ m

EXAMPLE 4 cont'd:

Summary of part (A) results:

	x meters downwind						
	25	50	100	200	600	1200	1549
Δh, meters	29	46	73	116	241	383	454

(B) Repeat the trajectory calculation in part (A) using a wind velocity of 6.5 ft/sec which is equivalent to 2 m/sec:

Inspection of equations (33) and (34) reveals that at any downwind distance x, the Δh for a windspeed of 2 m/sec is twice the Δh for a windspeed of 4 m/sec. Thus, the summary for this part (B) is:

	x meters downwind						
	25	50	100	200	600	1200	1549
Δh, meters	58	92	146	232	482	766	908

A plot of the trajectories obtained in parts (A) and (B) is presented in Figure 25.

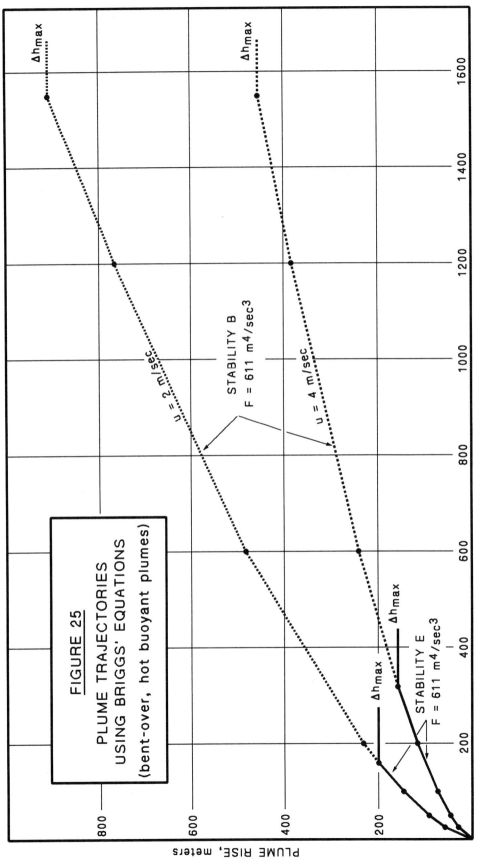

OTHER BRIGGS EQUATIONS

This chapter's primary focus has been on the Briggs plume rise equations for bent-over, hot buoyant plumes. However, Briggs[28] has also proposed equations for:

- the maximum rise of vertical, hot buoyant plumes in stable atmospheric conditions
- the maximum rise of vertical, cold jet plumes in stable atmospheric conditions
- the first and transitional rise stages as well as the maximum rise of bent-over, cold jet plumes in unstable or neutral atmospheric conditions
- the maximum rise of bent-over, cold jet plumes in stable atmospheric conditions

Vertical, hot buoyant plumes for atmospheric conditions of Pasquill stability classes E, F

(42) $\Delta h_{max} = 5.0\, F^{1/4}\, s^{-3/8}$ with dimensional units the same as on page 73

Vertical, cold jet plumes for atmospheric conditions of Pasquill stability classes E, F

Briggs proposed this equation for vertical jet plumes in calm conditions:

$$\Delta h_{max} = 4.0\, (F_m/s)^{1/4}$$

When combined with equation (32) defining the momentum flux parameter, F_m, the following equation is obtained:

(43) $\Delta h_{max} = 4.0\, (T_a/T_s)^{1/4}\, s^{-1/4}\, (rv_s)^{1/2}$

Bent-over, cold jet plumes for atmospheric conditions of Pasquill stability classes A, B, C, D

Briggs proposed this equation for the first and the transitional rise stages of bent-over, cold jet plumes in unstable or neutral conditions:

$$\Delta h = 2.3\, F_m^{1/3}\, u^{-2/3}\, x^{1/3}$$

When combined with equation (32) defining the momentum flux parameter, F_m, the following equation is obtained:

(44) $\Delta h = 2.3\, (T_a/T_s)^{1/3}\, (rv_s/u)^{2/3}\, x^{1/3}$

Briggs also proposed this equation for the maximum rise of bent-over, cold jet plumes in unstable or neutral conditions:

(45) $\Delta h_{max} = 6\, rv_s/u$

Bent-over, cold jet plumes for atmospheric conditions of Pasquill stability classes E, F

Briggs proposed this equation for the maximum rise of bent-over, cold jet plumes in stable conditions:

$$\Delta h_{max} = 1.5\, (F_m/u)^{1/3}\, s^{-1/6}$$

When combined with equation (32) defining the momentum flux parameter, F_m, the following equation is obtained:

(46) $\quad \Delta h_{max} = 1.5 \, (T_a/T_s)^{1/3} \, (rv_s)^{2/3} \, u^{-1/3} \, s^{-1/6}$

BRIGGS' 1975 LECTURE[†]

A number of international conferences and workshops have dealt with the theoretical aspects of turbulent diffusion in the atmosphere, dispersion modelling, and plume rise predictions. At one such workshop in 1975, Briggs presented a lecture on plume rise predictions.[23] The lecture was very broad in scope and attempted to provide plume rise equations for every conceivable plume type in every conceivable set of meteorological conditions. The presentation of the theoretical physics involved extremely complex mathematical concepts far beyond the understanding of any reader without a strong background in the advanced mathematics of the physicist. In fact, Briggs himself states that "some (readers) will be inclined to pass over the equations in this section".

In any event, Briggs' 1975 lecture appears to be a re-examination and a re-statement of much of his previous published work. It also appears to qualify many of his previously developed equations by defining some new constraints and adding some new formulations. However, it is difficult to explicitly summarize his 1975 lecture in any more exact terms. Until such time as that can be done, it is suggested that the Briggs plume rise equations continue to be defined as summarized on pages 72 and 73 herein (for bent-over, hot buoyant plumes) and on pages 81 and 82 (for other plume types). The only revision to those equations that might be considered as a result of the 1975 lecture is that Briggs appears to be recommending that the numerical coefficient in equation (41) be changed to 2.6 ... which seems to be a compromise between his 1969 value of 2.9 and his 1972 value of 2.4. In this book, his 1972 value of 2.4 will be retained.

[†] The post-1975 technical literature has many other publications by Briggs, including a 1984 publication on plume buoyancy effects.[38] Although this book does not review Briggs' work beyond his 1975 lecture, the reader may wish to do so.

Chapter 5

TIME-AVERAGING OF CONCENTRATIONS

THE COMPONENTS OF TURBULENCE

The primary factor in the dispersion of stack gas plumes is the intensity of atmospheric turbulence. Chapter 1 discusses the Pasquill stability classes which divide the total spectrum of atmospheric turbulence into six categories, A through F, within these limits:

Class A -- the most unstable class having the highest intensity of turbulence

Class F -- the most stable class having the lowest intensity of turbulence

The Pasquill classes are categorized either by windspeed and insolation, or by ambient temperature gradients (see Table 1). Since atmospheric turbulence can be visualized as consisting of seemingly random fluctuations (eddies) in wind velocity and direction, a more fundamental approach to categorizing turbulence can be gained by statistical techniques. An understanding of those techniques will provide an insight into the problem of converting plume dispersion estimates based on short-term data samples into estimates for longer time periods.

The prevailing wind velocity can be expressed as being the sum of instantaneous velocity vectors in the downwind, crosswind and vertical dimensions:

u = instantaneous velocity vector along the x-axis (coinciding with the mean downwind direction)

v = instantaneous velocity vector along the y-axis (crosswind to the mean downwind direction)

w = instantaneous velocity vector along the vertical z-axis

Each of the instantaneous velocity vectors can be expressed as having two components: a **mean (or average)** component and a **fluctuating (or deviation from the mean)** component. Thus:

$$u = \bar{u} + u'$$

$$v = \bar{v} + v'$$

$$w = \bar{w} + w'$$

where: $\bar{u}, \bar{v}, \bar{w}$ = mean values
u', v', w' = fluctuating values or deviations from the mean (plus or minus)

The fluctuating velocity components represent atmospheric eddy velocities, and their squares (u'^2, v'^2, and w'^2) represent eddy energies. The mean squares of the eddy velocities ($\overline{u'^2}$, $\overline{v'^2}$, and $\overline{w'^2}$) represent the kinetic energy of the eddies.

Chapter 5: Time-Averaging Of Concentrations

If we obtained measurements of the instantaneous x-axis velocity, u, taken during some specific time period, we could determine the mean velocity, $\bar{u}$, for that time period:

$$\bar{u} = N^{-1} \Sigma(u_1 + u_2 + u_3 \ldots + u_n) = (1/N) \Sigma u_i$$

where: N = number of measurements

We could also determine the deviation from that mean for each of the measurements:

$$u_i' = \bar{u} - u_i \qquad \text{where: } i = 1 \text{ through } N$$

= the fluctuating component, or eddy velocity component in the x-direction

We could then determine the mean square of the deviations, which is often called the "variance":

variance = mean square deviation

$$= (N-1)^{-1} \Sigma(u_i')^2 = (N-1)^{-1} \Sigma(\bar{u} - u_i)^2$$

$= \bar{u}'^2$ as conventionally written, which represents the eddy kinetic energy component in the x-direction

And, by definition, the square root of the mean square deviation is the standard deviation (σ_u) of the eddy velocity component in the x-direction:

$$\sigma_u = (\bar{u}'^2)^{1/2}$$

Finally, we can define the turbulence intensity vector i_x in the x-direction as the ratio of the standard deviation of the eddy velocity component to the mean velocity component in the x-direction:

$$i_x = (\bar{u}'^2)^{1/2} \div \bar{u} = \sigma_u \div \bar{u}$$

And similarly:

$$i_y = (\bar{v}'^2)^{1/2} \div \bar{u} = \sigma_v \div \bar{u}$$

$$i_z = (\bar{w}'^2)^{1/2} \div \bar{u} = \sigma_w \div \bar{u}$$

Thus, we have defined turbulence intensity in terms of statistical characteristics which could be measured and used to categorize turbulence classes. However, these characteristics would be more difficult to determine than those used for determining the Pasquill turbulence classes.

Assuming a mean horizontal wind velocity with no mean vertical velocity and no mean crosswind vector, then $\bar{w} = 0$ and $\bar{v} = 0$. Therefore, the instantaneous wind velocity can be expressed as:

(a) $\qquad u = \bar{u} + u' + v' + w'$

Equation (a) can be depicted in the x and y dimensions as a vector diagram to obtain the deviation of the horizontal wind direction (i.e., the angle θ'), as follows:

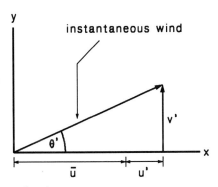

From the vector diagram, we obtain:

$$\tan \theta' = v'/(\bar{u} + u') \approx \theta' \qquad \text{where: } \theta' = \text{lateral wind deviation in radians}$$

And the mean square deviation of the lateral wind would be:

$$\overline{\theta'^2} = \overline{v'^2/(\bar{u} + u')^2}$$

Taking u' as very small relative to $\bar{u}$, we obtain:

$$\overline{\theta'^2} = \overline{v'^2}/\bar{u}^2$$

or taking square roots and using the definitions on the previous page:

$$\sigma_\theta = \sigma_v \div \bar{u} = i_y$$

The standard deviation of the lateral wind, σ_θ, is another measure of atmospheric turbulence intensity which could also be used to categorize turbulence classes. In fact, Gifford and Slade[7] have suggested that the Pasquill classes and σ_θ are related approximately as below:

Pasquill class	σ_θ in degrees	
	Gifford	Slade
A	25	25-27
B	20	---
C	15	13-14
D	10	8-9
E	5	2-7
F	2.5	2.5

A great deal of experimental field work in the determining of dispersion coefficients σ_y and σ_z, as well as the determining of the standard deviations of the lateral and vertical wind directions σ_θ and σ_λ, have established that:

$$\sigma_y \text{ is proportional to } \sigma_\theta x, \text{ and } \sigma_z \text{ is proportional to } \sigma_\lambda x$$

Chapter 5: Time-Averaging Of Concentrations

Tedious as it was, all of the foregoing discussion of wind velocity vectors, wind direction fluctuations and turbulence intensities was included for the express purpose of emphasizing as much as possible that:

(1) The dispersion coefficients σ_y and σ_z are related to fluctuations in the lateral and vertical wind directions which, in turn, are related to fluctuating eddy velocity components of the wind velocity vectors.

(2) The numerical values of i_x, i_y, i_z, σ_θ, σ_λ, σ_y and σ_z are <u>not</u> constants for any given set of meteorological conditions. They depend upon analyses of data obtained for specific sampling times (10 minutes to 1 hour) and for specific averaging times used in "smoothing" the data (in the order of 10 seconds).

To elaborate upon the second point just above, examine the continuous recording of wind velocities as presented in Figure 26. The upper section depicts continuous readings as recorded over some total sampling time T. To eliminate the inherent recording errors due to the inability of the velocity sensor and recorder to respond faithfully to high frequency fluctuations, the record is smoothed by averaging small time t increments of the continuous trace. The resulting smoothed data is depicted in the lower section of Figure 26.

The standard deviation of wind velocities obtained from the data in Figure 26 would increase with a longer sampling time T because more of the low frequency (long-period) fluctuations would be sensed over a longer sampling time. On the other hand, using larger increments of t for smoothing the data would decrease the standard deviation because it would dampen out the contribution of high-frequency (short-period) fluctuations.

Thus, the selection of sample time T and averaging time t affects the magnitude of the standard deviation obtained from the data. The design of the field experiments for obtaining such data must balance:

- The use of as large a sample time T as possible to maximize the inclusion of low frequency fluctuations without including gross changes in the overall meteorology, such as changes in the atmospheric stability, occurrence of inversions, rainstorms, etc.

- The use of an averaging time increment t[†] as small as possible so as to maximize the inclusion of high frequency fluctuations without violating the inherent limitations of the data gathering equipment.

[†] It must be kept in mind that the averaging time increment t used in the foregoing context refers to the time increment employed for smoothing the raw wind velocity data .... which is quite a different context than will be attributed to the words "averaging time" in the remainder of this chapter.

Chapter 5: Time-Averaging of Concentrations

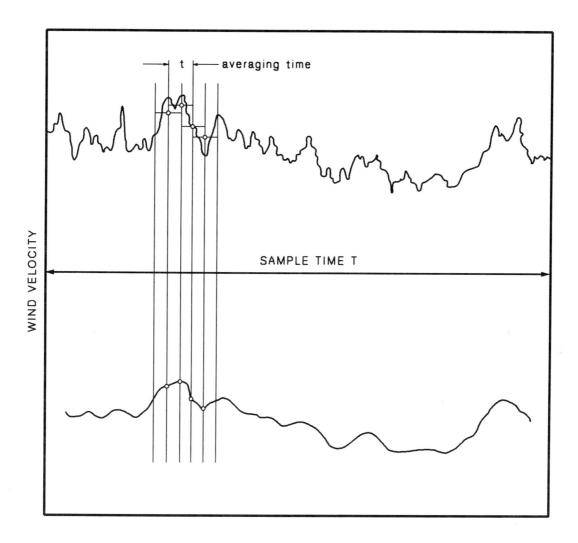

FIGURE 26

SMOOTHING A CONTINUOUS
WIND VELOCITY TRACE

TIME-AVERAGED DISPERSION COEFFICIENTS

Figure 27 represents a stack gas plume as it would appear instantaneously in a photographic snapshot, and as it would appear if averaged over a 10-minute or a 1-hour period as in photographic time exposures.

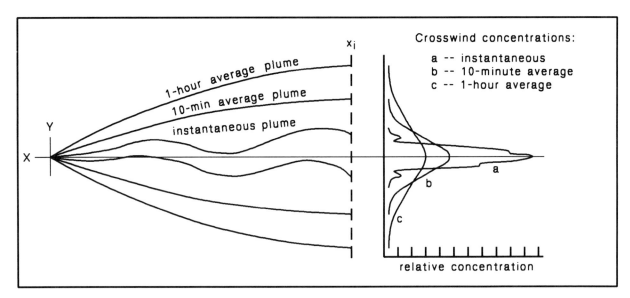

FIGURE 27

TIME-AVERAGED PLUMES

Assuming that the stack gas plume in Figure 27 is from a continuous emission source, we may think of the instantaneous snapshot plume as containing ΔQ emissions flowing from the source stack in Δt time (say 1 second). Thus, the amount of emissions passing point x_i in Figure 27 during any one second will be ΔQ. If the x-direction windspeed is constant and all of the emissions travel downwind at the same velocity as the wind, all that the human eye will see during any 1 second of time will be the instantaneous plume.

During a 10 minute period, 600 of the instantaneous 1-second plumes will travel from the source stack to point x_i. The 600 instantaneous plumes will travel in series (one behind the other) but will not necessarily follow identical paths. Variances in the wind velocity vectors will cause shifts in the wind direction. Some of the instantaneous plumes will travel to one side of the mean path, some will travel to the other side of the mean path, others may travel above or below the mean path, and all of them will meander to some extent. However, in the overall ensemble, the distribution of the 600 paths around the mean path will probably be Gaussian. If we took a photographic time-exposure over a 10-minute period, the photograph would see and record all 600 of the instantaneous plumes, which the human eye cannot do. The time-exposure picture is represented by the 10-minute average plume shown in Figure 27.

During a 1-hour period, 3600 of the instantaneous plumes will follow each other from the source stack to point x_i. During this longer time period, some larger variations in the travel paths will occur. If we took a photographic time-exposure over a 1-hour period, the photograph would see and record all 3600 of the instantaneous plumes. The 1-hour average plume would appear in the photographic time-exposure as shown in Figure 27.

Suppose we installed collectors on a crosswind line at point x_i, and in sufficient numbers to intercept all 600 of the instantaneous plumes passing point x_i in 10 minutes. Thus, during any 10-minute period, we could collect all of the emissions in the 600 plumes passing by in that time and we could obtain the spatial distribution of some measurable plume component such as SO_2 or perhaps radioactive tracers. From such experiments, σ_y and σ_z values could be derived for the specific meteorological conditions and other physical factors involved (such as windspeed, atmospheric stability, emission release height, emission rate, etc.). But it is very important to recognize that those σ values would be derived from data obtained over a 10-minute sample collection period and, therefore, would be representative of an averaging time of 10 minutes.

Now, let us suppose that the experimental determination of the σ_y and σ_z values had been based on a 1-hour sample collection period, with the appropriate increase in the number and location of collectors. The σ values so determined would then be representative of an averaging time of 1 hour.

This now brings us to the central question of the averaging time represented by the widely used Pasquill dispersion coefficients presented in Figures 17 and 18 of Chapter 3 herein. In other words, when those Pasquill dispersion coefficients are used in the Gaussian dispersion equations derived in Chapter 2, just what do the plume concentrations calculated by those equations represent? Are they 10-minute, 15-minute or 1-hour average concentrations? Or what averaging time do they represent?

There is no consensus of opinion as to the averaging time represented by the Pasquill dispersion coefficients. Slade's excellent review[7] of the many reported experimental determinations of σ_y and σ_z indicates that the experimental sampling times may have ranged from 3 minutes to 1 or 2 hours. Many respected practitioners in the field of air dispersion modelling assign a 10 to 15-minute averaging time to the Pasquill dispersion coefficients. The TVA appears to assign a 5-minute averaging time[39] to their dispersion coefficients. The State of Texas is reported[40] to use the Pasquill coefficients as representing a 10-minute averaging time. Turner's workbook[9] says that the Pasquill coefficients represent 10-minute averages. Hanna and Drivas[41] also state that the Pasquill coefficients are representative of 10-minute averages. However, many of the EPA's dispersion models ascribe an averaging time of 1-hour to the Pasquill coefficients. Therefore, an EPA dispersion model used to determine whether a specific plume will result in ground-level concentrations in violation of some specific air quality standard or regulation assigns a 1-hour averaging time to the Pasquill coefficients ... despite the fact that the EPA does not appear to have published a rationale for their decision to assign a 1-hour averaging time to the Pasquill dispersion coefficients. Apropos of this point, Hanna and Drivas[41] observe that "... it must be emphasized that very few (dispersion) models can account for variations in these (sampling and averaging) times. Usually these times are implicitly built into the empirical model with little or no discussion.".

EXTRAPOLATING TIME-AVERAGED CONCENTRATIONS

The concentration of stack gas emissions at a given receptor location can be calculated by using the Gaussian dispersion equations and the Pasquill dispersion coefficients as presented in Chapters 2 and 3 herein. Let us denote such a calculated concentration as C_p and define it as a time-averaged value which is valid for the sampling time period t_p over which the Pasquill coefficients have been experimentally determined. As discussed just above, that time period t_p appears to be within the range of 10 minutes to 1 hour, but there is no

Chapter 5: Time-Averaging Of Concentrations

conclusive consensus as to the exact value.

Ambient air quality standards setting forth maximum allowable ground-level concentrations of certain emission components (SO_2, NO_x, etc.) have been established by the EPA and by state regulatory agencies. Many of those standards are expressed in terms of averaging times such as 1-hour, 3-hours, 24-hours and 1-year.

The problem to be explored now is how to convert a time-averaged concentration C_p (valid for averaging time t_p) to an equivalent time-averaged concentration C_x (valid for averaging time t_x) for comparison to an air quality standard expressed in terms of averaging time t_x. In other words:

- Assume that we calculated the ground-level concentrations (resulting from a specific stack gas emission source) for a large number of reasonably probable wind velocities, atmospheric stabilities, and receptor locations. The maximum ground-level concentration C_p predicted by the calculations is found to be 100 $\mu g/m^3$ which we believe is a time-averaged value valid for an averaging time of say 30 minutes.

- Assume that the applicable ambient air standard for the specific emission component of concern is 75 $\mu g/m^3$ as a maximum 3-hour standard.

- How does one convert the predicted maximum ground-level concentration C_p of 100 $\mu g/m^3$, taken as valid for a 30-minute averaging time, to a 3-hour average C_x so that we can determine if the predicted ground-level emission concentration will violate the ambient air standard of 75 $\mu g/m^3$?

To approach an answer to the above question, let us examine Figure 27 again. It indicates that the crosswind dispersion coefficient σ_y increases with increased averaging time, since the crosswind concentration profile becomes broader and flatter. But what of the vertical dispersion coefficient σ_z? Slade[7] states:

"The standard deviation of the vertical wind direction (σ_z) will show <u>very little</u> increase for very long sampling times. This is in sharp contrast to the standard deviation of the horizontal wind direction (σ_y), which increases continuously with sampling time."

From this, we can infer that the vertical dispersion coefficient σ_z remains essentially constant with increases in the averaging time. Thus, if σ_y is directly proportional to the averaging time and σ_z is essentially constant or independent of any changes in the averaging time, then the ground-level **centerline** concentrations obtained with the Gaussian dispersion equation (15), derived in Chapter 2, are inversely proportional to the averaging time. That relationship can be expressed as:

(47) $\quad C_x/C_p$ = a function of (t_p/t_x)

where: t_p, t_x = any two averaging times

C_p, C_x = corresponding ground-level centerline concentrations

Many workers in this field have suggested the functional relation in equation (47) is:

(48) $\quad C_x/C_p = (t_p/t_x)^n$

Some have also suggested a single-value n in the range of 0.16 to 0.25:

Stewart, Gale, Crooks[7]	n = 0.20	
Hilst[7]	n = 0.25	
Wipperman[7]	n = 0.18	
Turner[9]	n = 0.17 - 0.20	
Nonhebel[42]	n = 0.16	

Others have suggested values of n to be a function of the atmospheric stability class:

Singer[7]	n = 0.30	stable
	= 0.60	unstable
EPA[43]	n = 0.32	class D
	= 0.52	classes B and C
	= 0.65	class A
State of Texas[40]	n = 0.18	classes E and F
	= 0.30	class D
	= 0.43	class C
	= 0.55	class B
	= 0.68	class A

All of the above suggested values of n for use in equation (48) can be summarized as:

-- a single-value n of about 0.20

-- values of n ranging from about 0.20 for stability classes E and F to 0.68 for stability class A

A very comprehensive field experimentation program to determine C_p/C_x ratios has been published by the TVA.[39] The program extended over two years and used a network of 14 monitors continuously recording ambient SO_2 concentrations downwind of a TVA power plant burning sulfur-containing fuel. Peak 5-minute averages (C_p) were determined for each instrument for each hour, each 2-hours, each 3-hours and each 24-hours during the test period. After reduction of the data, their C_p/C_x ratios were categorized in terms of percentile values ranging from 5 to 99 percent. Their set of 95th percentile C_p/C_x values have been fit by this equation:

(49) $\quad C_5/C_x = 0.98 + 0.00444\, t_x$

where: $\quad C_5$ = 5-minute average concentration

C_x = concentration over 1-hour, 2-hour or 3-hour averaging time

t_x = minutes

C_5/C_x = value equalled or exceeded by 95 percent of the pertinent C_5/C_x ratios

92   Chapter 5: Time-Averaging Of Concentrations

Equation (49) can be used to derive a more generalized expression which fits the TVA's published data quite well for any averaging time period up to 24 hours:

(50) $C_x/C_p = (220.2 + t_p)/(220.2 + t_x)$

where: C_x, C_p = concentrations over any two averaging times up to 24 hours
t_x, t_p = corresponding minutes
C_x/C_p = values equalled or not exceeded by 95 percent of the pertinent C_x/C_p ratios

Figure 28 compares the conversion of time-averaged concentrations within the range of 10 minutes to 3 hours using these three equations:

a -- equation (48) with n = 0.68
b -- equation (48) with n = 0.20
c -- equation (50) derived from the TVA data[39]

The three conversions a, b, and c are plotted twice in Figure 28:

- The upper section conversions a, b and c assume that the Pasquill dispersion coefficients represent a 1-hour averaging time and hence intersect C_x/C_p = 1 at t_p = 1 hour.
- The lower section conversions a, b and c assume that the Pasquill dispersion coefficients represent a 10-minute averaging time and thus intersect C_x/C_p = 1 at t_p = 10 minutes.

Figure 28 clearly shows that the EPA's assumption that Pasquill's σ values yield 1-hour average concentrations (see page 89) is biased toward the regulatory viewpoint, since that assumption leads to much higher converted concentrations than would result from assuming that Pasquill's σ values yield 10-minute average concentrations. Figure 28 also shows that equation (50) is less sensitive to the choice of base time t_p than are the other two equations. In other words, the converted 3-hour concentrations yielded by equation (50), assuming t_p to be either 1-hour or 10-minutes, vary by only 21 percent. By contrast, the same converted 3-hour concentrations yielded by equation (48), with n = 0.20, vary by 48 percent.

Equation (50), derived from the TVA data[39], is recommended for conversion of time-averaging periods within the range of 10 minutes to 24 hours. C_x/C_p values determined by equation (50) are summarized below in Table 10:

TABLE 10: CONVERSION OF TIME-AVERAGE CONCENTRATIONS
(within the range of 10 minutes to 24 hours)

t_x	Conversion Ratios, C_x/C_p	
	assuming t_p = 10 minutes	assuming t_p = 1 hour
10 minutes	1.00	1.22
1 hour	0.82	1.00
3 hours	0.58	0.70
8 hours	0.33	0.40
24 hours	0.14	0.17

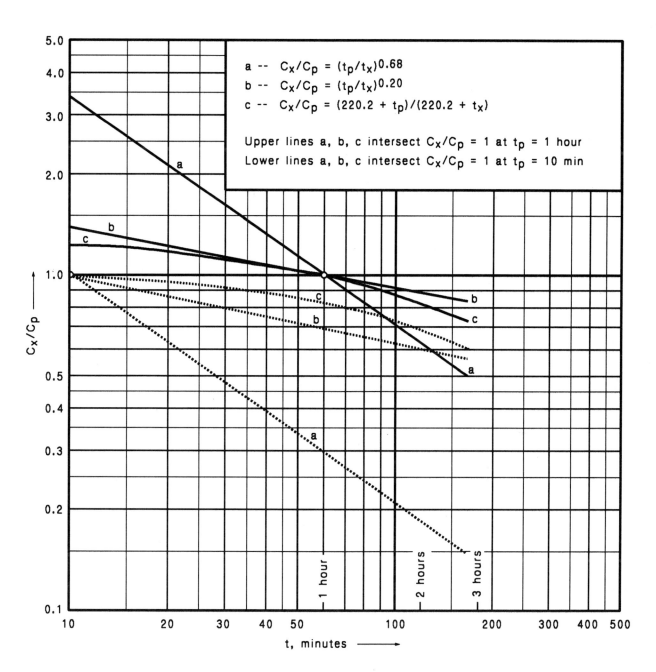

FIGURE 28

VARIOUS METHODS FOR CONVERTING TIME-AVERAGED CONCENTRATIONS

The TVA data[39] for longer time-averaging periods of from 24 hours to 1 year can be represented by the following equation:

(51) $C_x/C_p = (1/t_x)^{0.53}$ where: t_x = days
C_p = 24-hour average concentration

C_x/C_p values determined by equation (51) are summarized below in Table 11:

TABLE 11: CONVERSION OF TIME-AVERAGE CONCENTRATIONS
(within the range of 1 day to 365 days)

t_x	Conversion Ratios, C_x/C_p
1 day	1.00
7 days	0.36
30 days	0.16
365 days	0.044

When the appropriate meteorological data are available, computer models using the Gaussian dispersion equations can calculate 24-hour average concentrations which should be more accurate than using any of the above conversion equations. The data needed are hourly measurements in their actual sequence of occurrence over a period of at least one year. Thus, at least 8,760 hourly meteorological measurements are needed and must include this information for each hour:

- Ambient temperature
- Potential temperature gradient, $d\theta/dz$
- Wind velocity and direction
- Pasquill stability class

A polar receptor grid is selected, as shown in Figure 29, which divides the receptor area into wind direction sectors. The sectors are further divided into a number of receptor distances. The grid in Figure 29 locates 5 receptors in each of 16 sectors for a total of 80 receptors. The computer model calculates the hourly concentrations at each receptor as they would occur using the hourly meteorological data in their chronological order, assuming that the Pasquill dispersion coefficients in the Gaussian dispersion equations yield 1-hour average concentrations. In progressing through the hourly data, the computer would make 43,800 calculations of ground-level plume centerline concentrations (8,760 hours with 5 receptor concentration calculations for each hour) plus crosswind concentrations at increments of 22.5 degrees to both sides of the plume centerline.

For each day, the computer would make 120 calculations of ground-level plume centerline concentrations plus the requisite crosswind concentrations. The hourly concentrations values thus obtained for each receptor during each day are summed up and divided by 24 to determine 24-hour averages for each receptor for each day. For example:

- Figure 29 and Table 12 depict a day for which the wind direction is in sector 1 for 6 hours, sector 2 for 6 hours, sector 10 for 7 hours and sector 11 for 5 hours.

- Table 12 presents an array of the calculated hourly concentrations. For the sake of simplicity, crosswind concentrations outside of the 4 pertinent sectors are excluded. Thus, Table 12 has 120 centerline concentrations and 120 crosswind concentrations. The 24-hour averages for each of the 20 receptors involved are shown and the highest 24-hour averages (25 $\mu g/m^3$) occurred at receptors 48 and 49 in sector 10, which contained the wind for 7 hours of that day.

- The computer model would output 365 arrays such as shown in Table 12. The day or days exhibiting the highest 24-hour averages at any receptor could then be identified.

- Since it is conventional to depict polar grids, such as in Figure 29, in terms of the **direction from which the wind is coming**, sector 10 actually radiates to the northeast of the source stack and sector 2 actually radiates to the southwest of the source stack.

- The EPA's basic assumption that Gaussian dispersion calculations yield 1-hour average concentrations can be changed if desired. For example, if it is desired to assume that the calculations yield 10-minute averages rather than 1-hour averages, then Table 10 provides the C_x/C_p ratio of 0.82 by which the calculated 10-minute average results should be multiplied to yield 1-hour averages.

The EPA's computer dispersion model CRS-1 functions essentially as described just above except that it uses 180 receptors (36 sectors of 10 degrees each with 5 receptors in each sector).

Chapter 5: Time-Averaging Of Concentrations

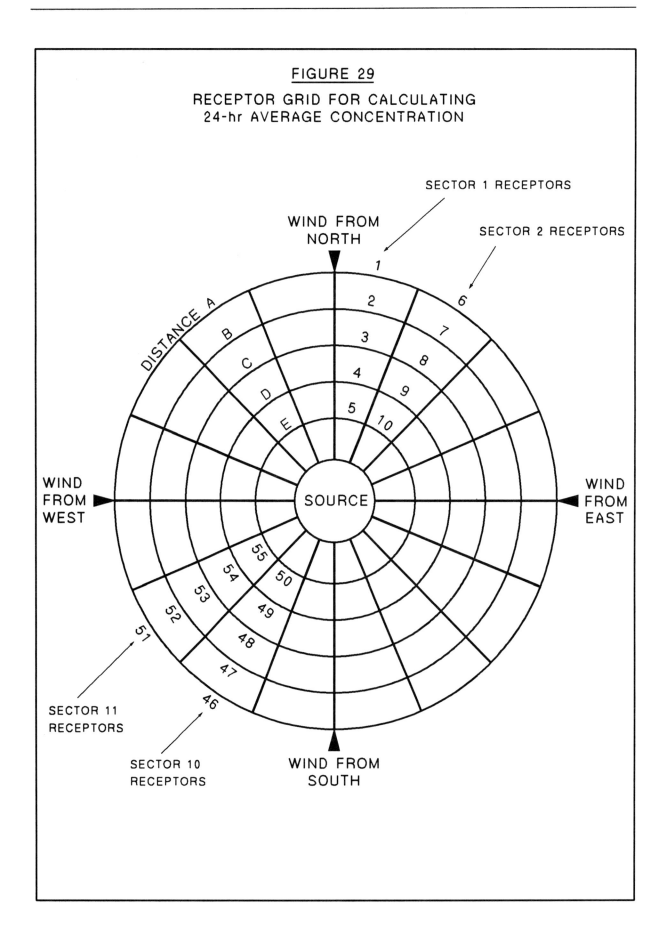

TABLE 12

EXAMPLE ARRAY OF 24-hr CONCENTRATIONS
(for one specific day)

Time	Concentrations in $\mu g/m^3$ at these receptors:																			
	1	2	3	4	5	6	7	8	9	10	46	47	48	49	50	51	52	53	54	55
0100	50	60	70	50	40	5	6	7	5	4										
0200	60	70	80	70	60	6	7	8	7	6										
0300	60	70	80	70	60	6	7	8	7	6										
0400	70	80	90	80	70	7	8	9	8	7										
0500	70	80	70	60	50	7	8	7	6	5										
0600	60	70	60	50	40	6	7	6	5	4										
0700	3	4	5	4	3	30	40	50	40	30										
0800	3	4	5	6	5	30	40	50	60	50										
0900	4	5	6	5	4	40	50	60	50	40										
1000	5	6	7	6	5	50	60	70	60	50										
1100	5	6	7	8	7	50	60	70	80	70										
1200	5	6	7	7	6	50	60	70	70	60										
1300											80	90	90	80	70	8	9	9	8	7
1400											70	80	90	80	70	7	8	9	8	7
1500											70	80	90	80	70	7	8	9	8	7
1600											60	70	80	90	80	6	7	8	9	8
1700											50	60	70	80	70	5	6	7	8	7
1800											50	60	70	70	60	5	6	7	7	6
1900											60	70	80	80	70	6	7	8	8	7
2000											6	7	8	8	7	60	70	80	80	70
2100											5	6	7	7	6	50	60	70	70	60
2200											5	6	7	8	7	50	60	70	80	70
2300											6	7	8	9	8	60	70	80	90	80
2400											7	8	9	8	7	70	80	90	80	70
24-hr avg	16	19	20	17	15	12	15	17	17	14	20	23	25	25	22	14	16	19	19	17

Concentrations at each receptor for each hour are calculated values obtained by using the Gaussian dispersion equations and the Briggs plume rise equations, and are based on the assumption that the Pasquill dispersion coefficients are valid for 1-hour averages. The ambient temperature, wind velocity, potential temperature gradient and atmospheric stability class were specified for each hour.

Chapter 6

WIND VELOCITY PROFILES

THE NEED FOR WIND VELOCITY PROFILES

Most of the plumes from industrial stacks are hot and buoyant. Since completely calm, non-windy conditions are relatively infrequent occurrences, the rise and dispersion of most industrial stack gas plumes will involve some wind and be characterized as the bent-over, hot buoyant plumes discussed in Chapter 4. One of the key parameters in determining the rise trajectory of bent-over, hot buoyant plumes is the wind velocity (see Briggs' equations in Chapter 4).

The wind velocity is also an important parameter in determining the vertical and crosswind Gaussian dispersion of stack gas components from continuous, point-source plumes (see the derivations in Chapter 2).

The wind velocity used in Briggs' equations for bent-over, hot buoyant plumes should be the velocity which prevails throughout the rise trajectory of the plume. Since the height of the plume rise along its trajectory will vary with the downwind distance from the source stack (see Figure 22 in Chapter 4), and since the wind velocity typically varies with altitude, the precise calculation of the Briggs plume rise height at any given downwind distance would require a reiterative trial-and-error estimation of the wind velocity at the average plume height between the plume source stack and the given downwind distance. Such precision is not usually justified in most cases. **However, as a minimum requirement, the wind velocity used in the Briggs equations should at least be determined at the height corresponding to the source stack exit.**

Similarly, the wind velocity used in the Gaussian dispersion equations should be the velocity which prevails throughout the vertical and crosswind spread of the plume. This creates a dilemma in that (a) the derivation of the generalized Gaussian dispersion equation assumes a constant wind velocity throughout the plume, and (b) the wind velocity is not constant throughout the plume since it typically varies with altitude. **Thus, as a minimum requirement, the wind velocity used in the Gaussian dispersion equation should be determined at the emission height (or the so-called "effective stack height") H_e for a given downwind distance**, where H_e is defined by:

(19) H_e = source height + plume rise

In most situations involving the calculation of stack gas dispersion, the available wind velocity data will have been determined at "ground stations" of about 10 meters elevation. Thus, there is a need for a generalized method of converting ground station wind velocities into profiles of velocity versus altitude so as to obtain:

- the wind velocity at the stack exit height for use in the Briggs plume rise equations, and

- the wind velocity at the emissions height H_e for use in the various Gaussian dispersion equations.

WIND VELOCITY VERSUS ALTITUDE

The winds aloft generally have a higher velocity than the winds at ground level. In other words, at any given time and place, wind velocity usually increases with altitude. The effect of altitude upon the velocity of the gradient winds (i.e., winds aloft) involves these two factors:

- the degree of turbulent mixing prevailing in the atmosphere at the given time, as characterized by the Pasquill stability class

- the surface area roughness, which induces surface friction at the given place

It has generally been agreed that the effect of altitude on the wind velocity is logarithmic and can be expressed as:

(52) $\quad u_z/u_g = (h_z/h_g)^n$

where: u_z = wind velocity at height z
u_g = wind velocity at ground station height
h_z = height z
h_g = ground station height (usually 10 m)

The EPA[5] uses the following exponent n values, as a function of the Pasquill stability class, in their Climatological Dispersion Model and ascribes the values to the work of DeMarrais[44]:

TABLE 13: EXPONENTS FOR EQUATION (52)

Stability class	Exponent n
A	0.10
B	0.15
C	0.20
D	0.25
E	0.25
F	0.30

Turner's workbook[9] presents data ascribed to Davenport[45], which yields the following values of the exponent n, as a function of the surface area roughness, for use in equation (52):

Level country: n = 0.14 - 0.15
Suburbs: n = 0.28
Urban areas: n = 0.41

Figure 30 illustrates how wind velocity increases with increasing altitude, and delineates the effects of atmospheric stability and of surface area roughness upon the altitude-to-wind velocity relationship, based upon the above exponent n values. Figure 30 indicates that:

- High turbulence and mixing (atmospheric stability class A) result in a much smaller increase of wind velocity with increasing altitude ... as compared to low turbulence (atmospheric stability class F).

- Level, smooth country areas also result in a smaller increase in wind velocity with increasing altitude ... as compared to urban areas with buildings which induce high surface friction.

It is recommended that the exponents used by the EPA's Climatological Dispersion Model (as listed in Table 13) be used in equation (52) to convert ground station wind velocities to velocities at the stack exit height and at the plume emissions height H_e for use in dispersion calculations. However, the effect of surface roughness should be kept in mind and perhaps applied in some judicious manner. For example, although the EPA's Climatological Dispersion Model[5] uses the exponents in Table 13 without any qualification as to the effect of surface roughness, their PAL[46] model uses the exponents in Table 14 and qualifies them as being applicable to a surface roughness typical of <u>urban</u> areas:

TABLE 14: EXPONENTS FOR EQUATION (52)
(for use in urban areas)

Stability class	Exponent n
A	0.15
B	0.15
C	0.20
D	0.25
E	0.40
F	0.60

Touma[47] published these exponent values based on averaging of field data gathered over 1-year periods for six different rolling terrain, <u>rural</u> sites in Missouri, Kansas, Iowa, Texas, Michigan and the Lake Erie area:

Stability Class	Exponent n Range	Average
A	0.10 - 0.14	0.11
B	0.09 - 0.18	0.12
C	0.08 - 0.17	0.12
D	0.12 - 0.21	0.17
E	0.20 - 0.33	0.29
F	0.41 - 0.56	0.45

The exponents in Table 13 have been recommended for use with equation (52) simply because they have found acceptance by dispersion modelling practitioners. However, **the reader may wish to use the PAL exponents in Table 14 for plumes in urban areas** ... or better yet, obtain site-specific data as recommended by Touma.[47]

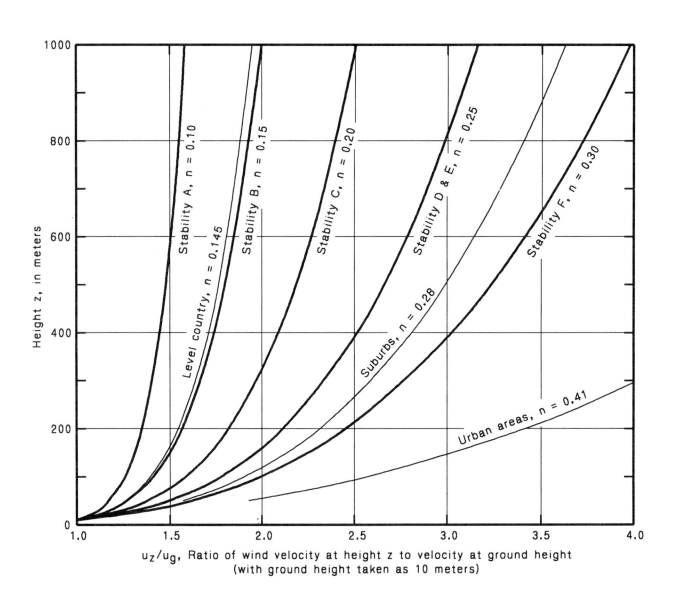

FIGURE 30: GENERALIZED EFFECT OF ALTITUDE ON WIND VELOCITY

Chapter 7

CALCULATING STACK GAS PLUME DISPERSION

INTRODUCTION

All of the component data and equations needed for calculating the dispersion of stack gas emissions have been presented in the preceding chapters of this book. The purpose of this chapter is to combine all of those component data and equations into a complete calculation procedure ... such as would be used in what is referred to as a "dispersion model".

The components of the calculation procedure or "dispersion model" are summarized below:

- The Gaussian dispersion equation for calculating the ground-level centerline and crosswind concentrations resulting from an elevated, continuous point-source plume. [see equation (14) in Chapter 2]

- The set of Briggs equations, as revised in 1972, for the rise of bent-over, hot buoyant plumes. [see recapitulation of Briggs' equations (33) through (41) in Chapter 4]

- The equations for calculating Briggs' buoyancy parameter F and stability parameter s. [see equations (29) and (31) in Chapter 4]

- Selection of the Pasquill atmospheric stability class as a function of a given windspeed and insolation, or as a function of a given ambient temperature gradient. [see Table 1 in Chapter 1]

- Selection of the potential temperature gradient consistent with the selected Pasquill stability class. [see Table 9 in Chapter 4]

- Selection of the dispersion coefficients to be used and the analytical equation for calculating the selected dispersion coefficients. The choices are:

  -- McMullen's equation for calculating Turner's version of Pasquill's rural dispersion coefficients. [see equation (27) and Table 6 in Chapter 3]

  -- Gifford's restatement of Briggs' urban dispersion coefficients based upon the work of McElroy and Pooler. [see equation (28) and Table 8 in Chapter 3]

- Selection of the windspeed conversion exponent consistent with the selected Pasquill stability class. [see equation (52) and Tables 13 and 14 in Chapter 6]

- Selection of the appropriate ratio for converting (or extrapolating) concentrations from one time-average basis to another. [see Tables 10 and 11, or equations (50) and (51) in Chapter 5]

In addition, the stack gas dispersion problem specifics must be available or selected. Those specifics include: stack height and diameter, stack exit temperature and gas flow rate, the concentration of pollutant species in the stack gas, and the ambient air temperature.

104   Chapter 7: Calculating Stack Gas Plume Dispersion

CALCULATION EXAMPLES

Examples 5, 6 and 7 illustrate the procedure for calculating the dispersion of a pollutant species, such as SO_2, from a stack gas plume so as to obtain the ground-level concentration of the pollutant at a receptor downwind of the source stack. The three examples were selected to show the effect upon dispersion of rural versus urban terrain, as well as the effect of a very unstable atmosphere (Pasquill stability class A) versus a very stable atmosphere (Pasquill stability class E).

The step-by-step procedure followed in examples 5, 6 and 7 is summarized briefly below:

- The emission rate of the pollutant species from the source stack is calculated in $\mu g/sec$ using the stack exit gas flow, the pollutant concentration within the stack gas, and the pollutant's molecular weight.
- The exit gas velocity from the stack is calculated in m/sec, using the stack exit gas flow and the stack exit diameter.
- Briggs' buoyancy and stability parameters (F and s) are calculated.
- The downwind distance to the point of maximum plume rise ($x_f = 3.5\ x^*$) is calculated using the appropriate Briggs equation.
- The surface wind velocity is converted to the equivalent velocity at the stack exit height, using the appropriate wind velocity conversion exponent.
- The plume rise at the point above the downwind receptor location is calculated with the appropriate Briggs equation and added to the stack's exit height, to obtain the emissions height or effective stack height (H_e) at that point.
- The surface wind velocity is converted to the equivalent velocity at the effective stack height, again using the appropriate wind velocity conversion exponent.
- The dispersion coefficients, σ_y and σ_z, for the given terrain (either rural or urban) are calculated at the receptor's downwind distance from the source stack.
- Finally, the Gaussian dispersion equation is used to calculate the ground-level concentration of the pollutant species, in $\mu g/m^3$, at the downwind receptor location.

Many of the parameters used in the above calculation procedure are dependent upon the selected atmospheric stability class. In other words, once the Pasquill stability class has been chosen, all of these selections must then be consistent with that choice:

- Potential temperature gradient
- Surface wind velocity
- Wind velocity conversion exponent
- The appropriate Briggs plume rise equation
- Dispersion coefficients (which are also dependent upon whether the terrain is rural or urban)

One point which should not be overlooked is that some of the Briggs plume rise equations involve the square and cube roots of the stability parameter s. Therefore, the sign of the potential temperature gradient must be discarded when calculating the stability parameter, and only its absolute value should be used. Otherwise, if the stability parameter is negative, the square and cube roots cannot be determined as a real number.

The ground-level concentrations determined in examples 5, 6 and 7 are short-term averages over a 10-minute to perhaps 1-hour period. Chapter 5 discusses how those results may be converted or extrapolated to longer time-averaging periods.

Chapter 7: Calculating Stack Gas Plume Dispersion

EXAMPLE 5: CALCULATION OF STACK GAS DISPERSION

Calculate the ground-level, centerline concentration of SO_2 at a receptor located 600 m downwind from a source stack.

GIVEN CONDITIONS AND PROBLEM SPECIFICS:

Type of terrain	Rural
Surface wind velocity	2 m/sec
Ambient temperature	15 °C
Pasquill stability class	A
Potential temperature gradient	-0.009 °K/m (see Table 9)
Source stack:	
Stack gas flow	19,132 Nm^3/hr[†]
Stack exit diameter	140 cm (or 1.4 m)
Stack exit height	76 m
Stack exit temperature	204 °C
SO_2 concentration in stack exit gas	1,428 ppm by volume
SO_2 molecular weight	64

Stack gas flow in kg-mols/sec

$\quad$ = [(19,132 Nm^3/hr)/(22.41 Nm^3/kg-mol)](1 hr/3600 sec) $\qquad$ = 0.237

SO_2 emission rate Q in kg-mols/sec

$\quad$ = (0.237)(1,428 × 10^{-6}) $\qquad$ = 338 × 10^{-6}

Stack emission rate Q in μg/sec

$\quad$ = (338 × 10^{-6})(64)(10^9 μg/kg) $\qquad$ = 21,630,000

Stack exit velocity v_s in m/sec

$\quad$ = (19,132/3600)[(204 + 273)/273]/(π)(1.4/2)^2 $\qquad$ = 6.032

Stability parameter s in sec^{-2}

$\quad$ = $(g/T_a)(d\theta/dz)$

$\quad$ = [9.807/(15 + 273)](0.009) $\qquad$ = 0.000306[††]

Buoyancy parameter F in m^4/sec^3

$\quad$ = $gv_s r^2(T_s - T_a)/T_s$

$\quad$ = (9.807)(6.032)(0.7)^2(477 - 288)/477 $\qquad$ = 11.49

[†] A Nm^3 is a normal cubic meter of gas measured at 0 °C and 1 atmosphere pressure. It is equivalent to 37.32 standard cubic feet of gas (i.e., SCF) measured at 60 °F and 1 atmosphere pressure.

$\qquad$ 1 kg-mol of any gas = 22.41 Nm^3 of the gas
$\qquad$ 1 lb-mol of any gas = 379 SCF of the gas.

[††] The sign of the potential temperature gradient $d\theta/dz$ indicates whether the atmosphere is stable or unstable. However, the absolute value of $d\theta/dz$ should be used when calculating the stability parameter s. The Briggs plume rise equations involve the square and cube roots of the stability parameter, which cannot be determined as a real number if the parameter is negative.

Since F < 55, then x_f

$= 3.5 \; x^* = 49 \; F^{0.625} = 49(11.49)^{0.625}$

= 225.4 m downwind distance to the point of maximum plume rise

Wind velocity conversion exponent n

= 0.10 for Pasquill stability class A (see Table 13)

The wind velocity u at stack exit height in m/sec

$= (2)(76/10)^{0.10}$ = 2.45

The downwind receptor distance x is specified above as 600 m. Thus, since $x > x_f$ and F < 55, the plume rise Δh_{max} at distance x

$= 21.4 \; F^{0.75} \; u^{-1}$ per equation (36)

$= 21.4(11.49)^{0.75}/2.45$ = 54.5 m

The effective stack height H_e in m

$= h_s + \Delta h = 76 + 54.5$ = 131

The wind velocity u at the effective stack height in m/sec

$= (2)(131/10)^{0.10}$ = 2.59

McMullen's constants in Table 6 for use in equation (27) to obtain rural dispersion coefficients for Pasquill stability class A:

    For σ_z: I = 6.035   J = 2.1097   K = 0.2770
    For σ_y: I = 5.357   J = 0.8828   K = -0.0076

Using equation (27):

$\sigma_z = \exp[\; I + J(\ln x) + K(\ln x)^2 \;]$

$ = \exp[6.035 + 2.1097(\ln 0.6) + 0.2770(\ln 0.6)^2]$ = 153 m at 0.6 km distance

$\sigma_y = \exp[5.357 + 0.8828(\ln 0.6) - 0.0076(\ln 0.6)^2]$ = 135 m at 0.6 km distance

Using the Gaussian dispersion equation (14) in Chapter 2 and recognizing that the problem only calls for the ground-level centerline concentration of SO_2:

$C = [Q/(u\sigma_z\sigma_y\pi)] \; \exp(-y^2/2\sigma_y^2) \; \exp(-H_e^2/2\sigma_z^2)$

$ = [21{,}630{,}000/(2.59)(153)(135)(\pi)] \; \exp(0) \; \exp[-0.5(131/153)^2]$ = 89.2 $\mu g/m^3$

Chapter 7: Calculating Stack Gas Plume Dispersion

EXAMPLE 6: CALCULATION OF STACK GAS DISPERSION

Repeat the calculation in example 5 for urban terrain rather than rural terrain.

THESE CALCULATIONS REMAIN THE SAME AS IN EXAMPLE 5:

Stack gas flow in kg-mols/sec
 = (19,132)/(22.41)](1/3600) = 0.237

Stack SO_2 emission rate Q in μg/sec
 = (0.237)(1,428 × 10^{-6})(64)(10^9) = 21,630,000

Stack exit velocity v_s in m/sec
 = (19,132/3600)(477/273)/(π)(1.4/2)^2 = 6.032

Stability parameter s in sec^{-2}
 = (9.807/288)(0.009) = 0.000306

Buoyancy parameter F in m^4/sec^3
 = (9.807)(6.032)(0.7)^2(477 - 288)/477 = 11.49

Since F < 55, then x_f
 = 49(11.49)$^{0.625}$ = 225.4 m

Urban wind velocity conversion exponent n
 = 0.15 for Pasquill stability class A

The wind velocity u at stack exit height in m/sec
 = (2)(76/10)$^{0.15}$ = 2.71

Since x > x_f and F < 55, the plume rise Δh_{max} at distance x
 = 21.4(11.49)$^{0.75}$/2.71 = 49.3 m

The effective stack height H_e in m
 = h_s + Δh = 76 + 49.3 = 125

The wind velocity u at the effective stack height in m/sec
 = (2)(125/10)$^{0.15}$ = 2.92

THE REMAINDER OF THE CALCULATIONS DIFFER FROM EXAMPLE 5:

Gifford's constants in Table 8 for use in equation (28) to obtain urban dispersion coefficients for Pasquill stability class A:

   For σ_z: L = 240   M = 1.00   N = 0.50
   For σ_y: L = 320   M = 0.40   N = -0.50

Using equation (28):

σ_z = Lx(1 + Mx)^N = (240)(0.6)[1 + 1.00(0.6)]$^{0.50}$ = 182 m at 0.6 km distance

σ_y = Lx(1 + Mx)^N = (320)(0.6)[1 + 0.40(0.6)]$^{-0.50}$ = 172 m at 0.6 km distance

And thus:

C = [Q/(u$\sigma_z\sigma_y\pi$)] exp(-y^2/2σ_y^2) exp(-H_e^2/2σ_z^2)

 = [21,630,000/(2.92)(182)(172)(π)] exp(0) exp[-0.5(125/182)^2] = 59.5 μg/m^3

108   Chapter 7: Calculating Stack Gas Plume Dispersion

EXAMPLE 7: CALCULATION OF STACK GAS DISPERSION

Repeat the calculation in example 5 using a downwind receptor distance of 1.50 km rather than 600 m and a Pasquill stability class E instead of class A.

THESE CALCULATIONS DIFFER FROM EXAMPLE 5:

Potential temperature gradient for stability class E   0.015 °K/m   (see Table 9)

Stability parameter s in sec^{-2}
= (9.807/288)(0.015)   = 0.00051

THESE CALCULATIONS REMAIN THE SAME AS IN EXAMPLE 5:

Stack gas flow in kg-mols/sec
= (19,132)/(22.41)](1/3600)   = 0.237

Stack SO_2 emission rate Q in μg/sec
= (0.237)(1,428 × 10^{-6})(64)(10^9)   = 21,630,000

Stack exit velocity v_s in m/sec
= (19,132/3600)(477/273)/(π)(1.4/2)^2 = 6.032

Buoyancy parameter F in m^4/sec^3
= (9.807)(6.032)(0.7)^2(477 - 288)/477   = 11.49

Since F < 55, then x_f
= 49(11.49)$^{0.625}$ = 225.4 m

THE REMAINDER OF THE CALCULATIONS DIFFER FROM EXAMPLE 5:

Wind velocity conversion exponent n
= 0.25 for Pasquill stability class E (see Table 13)

The wind velocity u at stack exit height in m\sec
= (2)(76/10)$^{0.25}$ = 3.32

Calculate 1.84 u s$^{-1/2}$ in m
= 1.84(3.32)(0.00051)$^{-1/2}$ = 271

Since 1.84 u s$^{-1/2}$ > x_f and x > x_f and F < 55, the plume rise at x
= 21.4 F$^{0.75}$ u^{-1} = 21.4(11.49)$^{0.75}$/3.32   = 40 m

The effective stack height H_e in m
= 76 + 40   = 116

The wind velocity u at the effective stack height in m/sec
= (2)(116/10)$^{0.25}$ = 3.69

McMullen's constants in Table 6 for use in equation (27) to obtain rural dispersion coefficients for Pasquill stability class E:

For σ_z: I = 3.057   J = 0.6794   K = -0.0450
For σ_y: I = 3.922   J = 0.9222   K = -0.0064

Using equation (27):

σ_z = exp[3.057 + 0.6794(ln 1.5) - 0.0450(ln 1.5)^2]   = 27.8 m at 0.6 km distance
σ_y = exp[3.922 + 0.9222(ln 1.5) - 0.0064(ln 1.5)^2]   = 73.3 m at 0.6 km distance

And thus:

C = [21,630,000/(3.69)(27.8)(73.3)(π)] exp[-0.5(116/27.8)^2]   = 0.15 μg/m^3

Chapter 7: Calculating Stack Gas Plume Dispersion

EXAMPLE GRAPHS OF GROUND-LEVEL CONCENTRATIONS VERSUS DOWNWIND RECEPTOR DISTANCES

Figures 31 and 32 show the general relationship between calculated ground-level pollutant concentrations and downwind receptor distances for a specific stack gas dispersion problem. Such graphs illustrate the effect of various parameters on the calculated ground-level concentrations. Figures 31 and 32 are based on these problem specifics and conditions:

Stack gas flow	19,132	Nm^3
Stack diameter	140	cm
Stack exit temperature	204	°C
Ambient temperature	15	°C
Stack height:		
For Figure 31	76	m
For Figure 32 (upper section)	76	m
For Figure 32 (lower section)	152	m
Stability class:		
For Figure 31	as noted	
For Figure 32	class A	
Dispersion coefficients:		
For Figure 31	as noted	
For Figure 32	rural	

These are the same problem specifics and conditions used in examples 5 and 6, where the ground-level SO_2 concentrations were calculated (for atmospheric stability A, 2 m/sec windspeed and 0.6 km downwind receptor distance) as 89.2 $\mu g/m^3$ using rural dispersion coefficients and 65.5 $\mu g/m^3$ using urban coefficients. The graphs for stability A and 2 m/sec windspeed in Figure 31 can be read at 0.6 km downwind to obtain 89 $\mu g/m^3$ using rural coefficients and 65 $\mu g/m^3$ using urban coefficients, which simply confirms that Figure 31 is indeed based on the same problem specifics as were examples 5 and 6.

The graphs in Figure 31 illustrate the effect on the calculated ground-level concentrations of:

- Varying the stability class with all other specifics held constant except for the wind velocity and the wind velocity conversion exponent, which must be selected to be appropriate for each stability class.

- Varying the dispersion coefficients from rural to urban values with all other specifics held constant.

- Varying the windspeed within each stability class, with all other specifics held constant.

Figure 32 illustrates the effect of varying source height within a given atmospheric stability class, with all other specifics held constant. As shown in Figure 32, each windspeed graph for a given stack height exhibits a maximum ground-level pollutant concentration at a specific downwind receptor distance ... which can be denoted as C_{max} for that windspeed. The windspeed graph with the highest C_{max} value, which can be denoted as $C_{max-max}$, is often called the "critical" windspeed or "u_{crit}" for the given stack height. In Figure 32:

- $C_{max-max}$ occurs at a u_{crit} of 0.8 m/sec for a 76 m stack height
- $C_{max-max}$ occurs at a u_{crit} of 0.4 m/sec for a 152 m stack height

110   Chapter 7:   Calculating Stack Gas Plume Dispersion

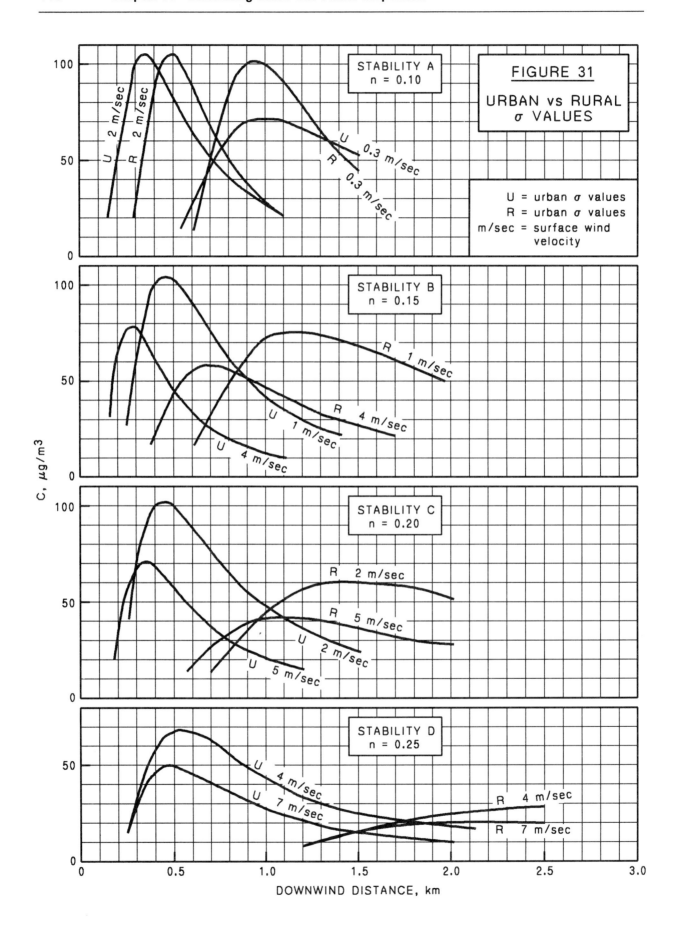

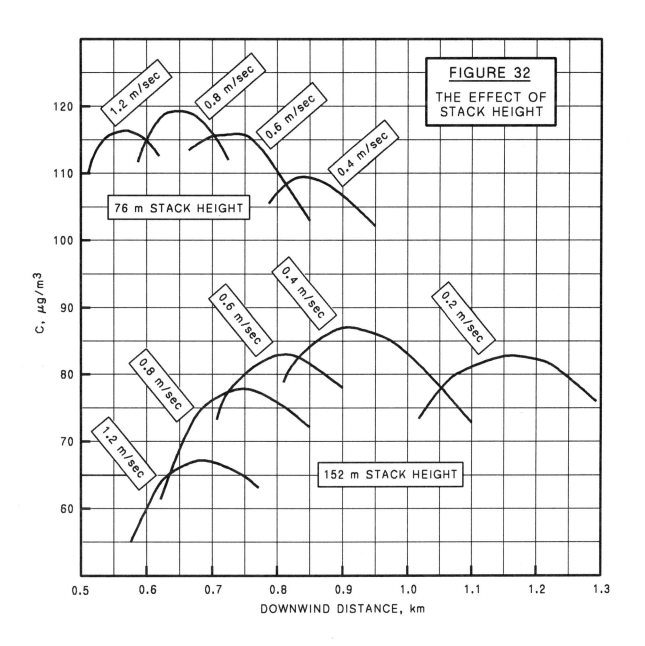

FIGURE 32
THE EFFECT OF STACK HEIGHT

Figures 31 and 32 are typical only to the extent that they illustrate the effects of varying certain problem specifics for a single selected stack gas dispersion problem. They cannot be generalized as applying to any specific dispersing plume. **There is no way to "outguess" the critical windspeed for any given stack gas dispersion problem.** Increasing the wind velocity term in the Gaussian dispersion equation yields lower ground-level concentrations. However, increasing the wind velocity term in the Briggs plume rise equations yields higher ground-level concentrations because higher wind velocities result in lower plume rises and, hence, lower effective stack heights.

Thus, there is no substitute for calculating the concentrations versus downwind distances for a set of wind velocities and plotting the results, as in Figure 32, to determine the critical windspeed for a specific stack gas dispersion problem. After having completed at least two such plots at two different wind velocities, Figure 32 can be quite useful in deciding whether the subsequent plots should be determined at higher or lower wind velocities. Similarly, Figure 31 can be quite useful in evaluating the effect of varying the atmospheric stability class or of using rural versus urban dispersion coefficients. To that extent, Figures 31 and 32 provide useful guidance ... but they are not intended to be used as substitutes for problem-specific calculations.

PLOTTING ISOPLETHS

A concentration isopleth is a line of constant ground-level pollutant concentration plotted upon a map or receptor location grid. The ground-level pollutant concentrations generated by a computerized Gaussian dispersion model can readily provide the data from which to plot concentration isopleths. Figure 33 illustrates how isopleths are derived from an array of calculated ground-level pollutant concentrations versus receptor locations:

- The centerline and crosswind ground-level pollutant concentrations of a dispersing stack gas plume are arrayed at their respective locations on the receptor grid.

- The isopleth lines of constant concentration (such as those for 75, 100 and 110 $\mu g/m^3$ in Figure 33) are determined by interpolation between the arrayed values. The interpolations are made between adjacent downwind concentration values as well as between adjacent crosswind values.

The isopleths in Figure 33 are based upon the same problem specifics as the 0.6 m/sec wind velocity graph in Figure 32, namely:

Stack gas flow	19,132	Nm^3
Stack diameter	140	cm
Stack height	76	m
Stack exit gas temperature	204	°C
Ambient temperature	15	°C
SO_2 in the stack exit gas	1,428	ppm by volume
Atmospheric stability class	A	
Dispersion coefficients	rural	

Figure 33 illustrates the short-time average ground-level concentration isopleths for a relatively simple, single point-source dispersion under constant conditions of wind direction, windspeed and atmospheric stability. Therefore, the isopleths exhibit a perfect symmetry about the dispersing plume's downwind centerline.

Isopleths derived from multiple sources, and those derived for annual or seasonal average concentrations, would not be symmetrical and would exhibit very irregular shapes such as presented in Figure 34. The longer-time average concentration isopleths reflect the occurrence of various wind directions and velocities as well as various atmospheric stability conditions during a longer-time period. In the case of irregular terrain situations, such as mountains or valleys, the isopleths become even more irregular in shape ... in much the same manner that elevation contours on a map become very irregular in those same types of terrain situations.

THE ACCURACY OF DISPERSION MODELS

This book devotes a great many pages to the detailed derivation of precise mathematical equations. The multitude of assumptions and constraints defined in those derivations tend to be overlooked and forgotten amongst the final equations and their applications in example problems. The book would be virtually unreadable if all of the assumptions and constraints were re-stated each time that an equation appears in the text. Thus, the reader may easily fall into the trap of confusing precision with accuracy. **Calculating plume ground-level concentrations with computers capable of providing results with 12-digits beyond the decimal point does not necessarily make the calculated concentrations accurate. Nor does the expression of relationships in precise algebraic form necessarily make those expressions accurate. Most of the assumptions and constraints discussed in this book can be, and many have been, expressed in algebraic terms ... but they are still no more than assumptions and constraints.** A careful re-reading of this book would reveal the long chain of assumptions and constraints leading to a very simple, single point-source stack gas dispersion model ... which is still a long way from the very complex, state-of-the-art dispersion models now in use.

Leaving aside the dispersion equations and examining the data base for many of the input specifics used in those equations (such as atmospheric stability classifications, dispersion coefficients, plume rise parameters, wind velocity profiles, etc.), we find that the data base leaves much to be desired. The validity of some of the input specifics used in the dispersion equations is perhaps even more difficult to rationalize than are the dispersion equations themselves.

This discussion of the shortcomings in dispersion models is neither unique nor profound. The literature abounds with similar discussions.[48-53] But unfortunately, there persists the belief that dispersion models calculate plume ground-level concentrations which are within a factor of three of the actual concentrations in the real world. That belief is given credence in some EPA dispersion models which print out this message: "Concentration estimates (from the dispersion model) may be expected to be within a factor of three for (certain conditions)". The only apparent justification for such optimism appears to be based on Pasquill's sensitivity study[1] of the Gaussian dispersion equation. Pasquill assumed that the source emission rate, wind velocity and stability class were known exactly, which is rarely the case. Pasquill also assumed that the emission height and dispersion coefficients were accurate within 15 percent, which is quite debatable. In any event, on the basis of those assumptions, Pasquill arrived at an overall sensitivity of 50 percent in the magnitude of the calculated short-time average concentration. As discussed in Chapter 5, we are not sure whether those short-time averages are 10-minute or 1-hour averages.

114 Chapter 7: Calculating Stack Gas Plume Dispersion

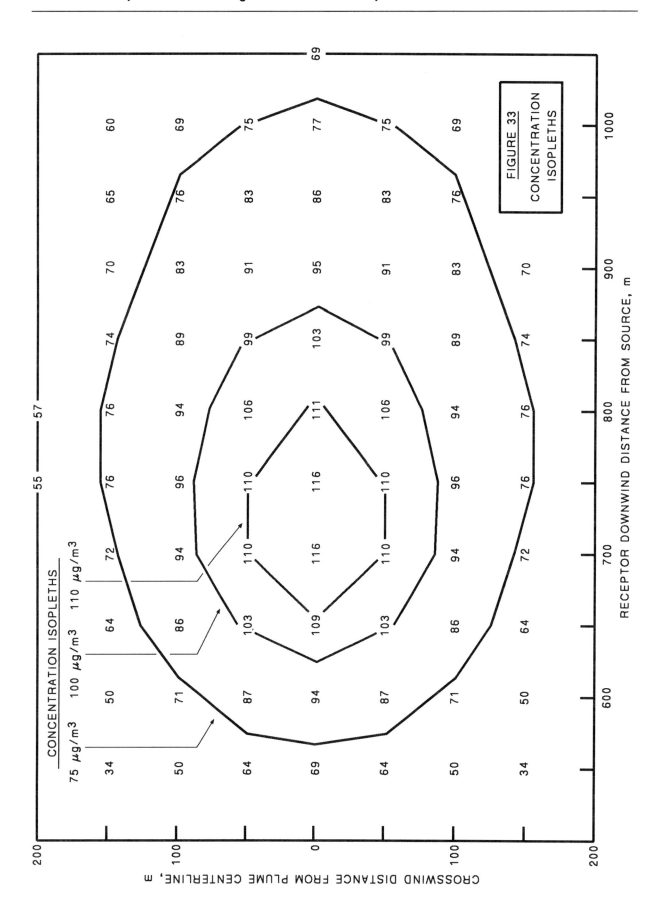

FIGURE 33
CONCENTRATION ISOPLETHS

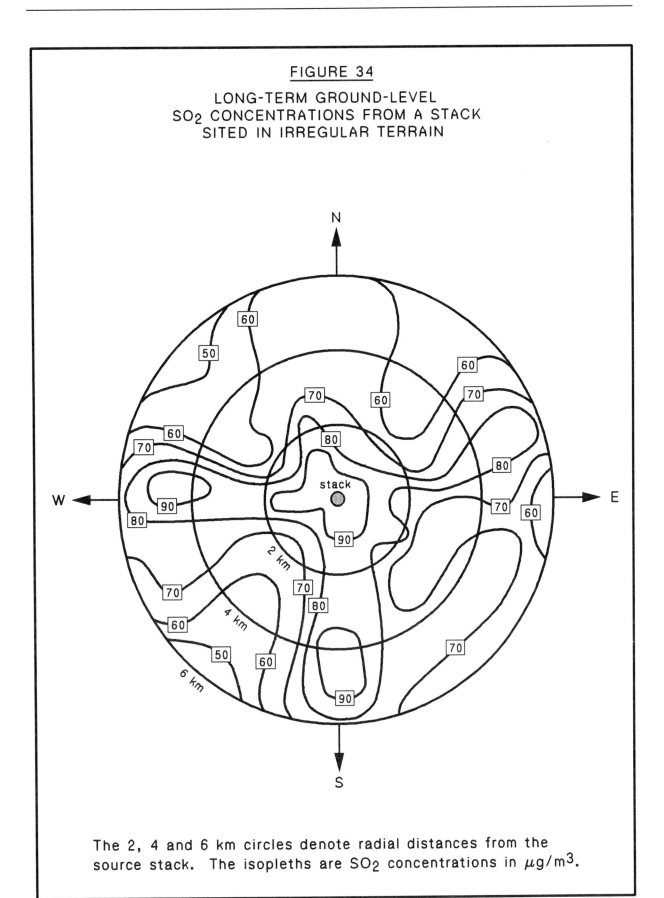

FIGURE 34

LONG-TERM GROUND-LEVEL SO$_2$ CONCENTRATIONS FROM A STACK SITED IN IRREGULAR TERRAIN

The 2, 4 and 6 km circles denote radial distances from the source stack. The isopleths are SO$_2$ concentrations in $\mu g/m^3$.

The key point is that Pasquill's sensitivity analysis **assumed** that the accuracy of the effective stack height (by plume rise calculations) and the accuracy of the dispersion coefficients were each within a 15 percent range, and it **also assumed** that the windspeed and the atmospheric stability class were known exactly. On that basis, the sensitivity analysis found that the ground-level concentrations calculated by the Gaussian dispersion equation varied by some 50 percent (i.e., a factor of two) if the effective stack height and the dispersion coefficients were varied by 15 percent. On a purely mathematical basis, Pasquill's sensitivity analysis is no doubt correct. But it should not be translated into a quantitative assessment of the accuracy of air dispersion models including all of their required input specifics. **We cannot state with any degree of certainty that the atmospheric stability class and the windspeed are known exactly at the plume height and, more importantly, that they are constant for the entire distance from the source stack to the downwind receptor.** The assumption of constant conditions along the entire path of a dispersing plume's path is fundamentally inherent in the derivation of the Gaussian dispersion equation (see Chapter 2). Whether or not such homogeneous conditions actually prevail for a specific real world dispersion situation is probably a matter of chance, particularly for large source-to-receptor distances.

Beychok published a sensitivity analysis of the Gaussian dispersion equation[53] which went into much more detail than did Pasquill's analysis. Briefly, Beychok's analysis included:

- A base model that used Briggs' plume rise equations, Pasquill's dispersion coefficients, a power-law conversion of the surface windspeed to obtain windspeeds at the stack height (for use in the plume rise equations) and at the plume centerline height (for use in the Gaussian dispersion equation), and the calculated ground-level concentrations were taken to be 1-hour average values as per the EPA's apparent position.

- A comparative "adjusted model" which decreased the vertical Pasquill dispersion coefficients (σ_z) by 25 percent and increased the calculated plume rises by 20 percent.

- A comparative "adjusted model plus wind shift correction" in which a 10° shift in wind direction was assumed to occur.

- A comparative "adjusted model plus wind shift and C_{10}/C_{60} corrections" in which an overprediction factor of 2.5 is taken into account because of the base model's assumption that the dispersion coefficients result in 1-hour average concentrations rather than 10-minute average concentrations.

The various models all used the same problem specifics: a stack height of 61 m, rural terrain dispersion coefficients, a surface windspeed of 2 m/sec, Pasquill stability class B, a receptor location 10 km downwind from the source stack, an exponent of 0.52 for use with equation (47) in determining the C_{10}/C_{60} correction, and an exponent of 0.15 for use with equation (52) in converting surface windspeeds to windspeeds aloft. It should be noted that a distance of 10 km to the receptor and a windspeed of 2 m/sec amounts to a plume travel time of almost one and a half hours, during which time all pertinent conditions must remain constant to be consistent with the derivation of the Gaussian dispersion equation.

There are few knowledgeable practitioners in the field of dispersion modelling who would dispute that plume rises calculated by the Briggs equations could easily over or underpredict by at least 20 percent. Likewise, there are few knowledgeable practitioners would dispute that the Pasquill dispersion coefficients could easily have an uncertainty range of plus or minus 25 percent. The assumed 10° shift in the wind direction certainly seems reasonable.

The vertical, ground-level centerline concentration profiles calculated for Beychok's base model and his first two comparative adjusted models were extracted from his sensitivity study[53] and are presented as profiles A, B and C in Figure 35. It is readily apparent from Figure 35 how much the base model A overpredicts the concentrations at various downwind distances compared to the two adjusted models B and C. As can be seen, the overprediction is more pronounced at downwind distances of up to 4 km than it is at distances between 4 and 10 km.

Table 15 (also extracted from Beychok's study) includes his third adjusted model which takes into account the overprediction resulting from assuming that the Pasquill dispersion coefficients result in 1-hour average concentrations rather than 10-minute average concentrations. As shown in Table 15, **the overprediction factor when comparing the base model with the final adjusted model varies from a factor of 6 at downwind distances of 6-10 km to a factor of 80 at a downwind distance of 2 km from the source stack ... as a result of seemingly minor changes in a few of the many variables involved and from assuming that the Gaussian models yield 10-minute averages rather than 1-hour averages.**

It is not the intent of this discussion of the accuracy of Gaussian dispersion models to denigrate the importance and the usefulness of such models. In actual fact, those models are the best tools we have. But we must keep in mind that they are only tools and they do not provide the ultimate truth. The models can be very useful in qualitatively predicting which industrial plant designs and which plant sites should result in a lower impact on air quality. They can also be very useful in evaluating and selecting emission control methods. **But they are far from being able to consistently predict actual dispersing plume concentrations within a factor of two or three.** The consistent prediction of actual plume concentrations within a factor as high as ten is probably a more realistic assessment of the accuracy of Gaussian dispersion models used to calculate and predict short-time average concentrations. Once again, as stated at the beginning of this section, we must not confuse precision with accuracy.

TABLE 15

1-HOUR GROUND-LEVEL CONCENTRATIONS ($\mu g/m^3$)
CALCULATED UNDER THE PLUME CENTERLINE

Receptor downwind distance (km)	(A) Base model	(B) Adjusted model	(C) Adjusted model plus wind shift	(D) Adjusted model plus wind shift and C_{10}/C_{60} corrections	Over-prediction ratio, (A)/(D)
2	16.1	1.1	0.5	0.2	80
3	24.3	9.2	4.4	1.7	14
4	20.5	13.8	6.0	2.4	9
5	15.8	13.7	5.8	2.3	7
6	12.0	12.0	4.7	1.9	6
7	9.4	10.1	3.9	1.5	6
8	7.4	8.4	3.1	1.2	6
9	6.0	7.0	2.5	1.0	6
10	5.0	5.9	2.0	0.8	6

FIGURE 35
SENSITIVITY STUDY OF GAUSSIAN DISPERSION MODELS [53]

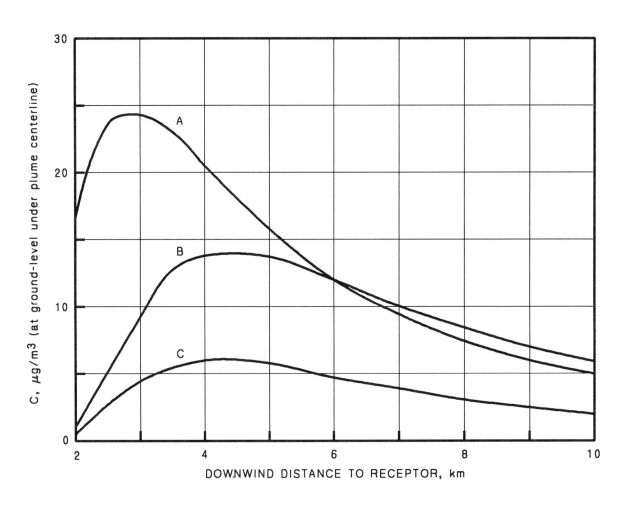

A Brigg's plume rise, Gaussian dispersion, stack and plume height windspeeds obtained by power-law conversion of surface windspeeds. Concentrations taken to be 1-hour averages per the EPA.

B Same as A except that calculated plume rises were increased by 20 percent and the vertical Pasquill dispersion coefficients were decreased by 25 percent.

C Same as B except that the ground-level concentrations reflect a 10° shift in the wind direction.

Chapter 8

TRAPPED PLUMES

INTRODUCTION

When a stack gas plume is located beneath an inversion aloft, emissions from the plume are prevented from dispersing upward through the bottom of the inversion aloft and the plume is described as being "trapped". The general behavior of a trapped plume was discussed briefly in Chapter 2.

The purpose of this chapter is to derive a generalized Gaussian dispersion equation which applies to either trapped or non-trapped plumes from a continuous point-source. The derivation of that equation will be based upon the concept of multiple reflections as originated by Hewson and his colleagues.[10,54] From the derivation of the generalized Gaussian dispersion equation (13) in Chapter 2, we can obtain the following equation that applies in the absence of any upward or downward barrier to vertical dispersion:

(53) $$C = \frac{Q}{u} \cdot \frac{f}{\sigma_y \sqrt{2\pi}} \cdot \frac{g_1}{\sigma_z \sqrt{2\pi}}$$

where: C, Q, u, σ_y and σ_z are all as defined in Chapter 2

f = crosswind dispersion parameter
  = $\exp[-y^2/(2\sigma_y^2)]$

g_1 = vertical dispersion parameter
  = $\exp[-(z - H)^2/(2\sigma_z^2)]$

$z = z_r$ = receptor location in the z-dimension, m

$H = H_e$ = emission height or effective stack height, m

When the ground is taken into consideration as a downward barrier to vertical dispersion, and the reflection of emissions which have diffused downward to the ground is included:

(54) $$C = \frac{Q}{u} \cdot \frac{f}{\sigma_y \sqrt{2\pi}} \cdot \frac{g_1 + g_2}{\sigma_z \sqrt{2\pi}}$$

where: g_2 = ground reflection vertical dispersion parameter
  = $\exp[-(z + H)^2/(2\sigma_z^2)]$

Equation (54) is completely equivalent to the generalized Gaussian dispersion equation (13) derived in Chapter 2. A comparison of equations (53) and (54) shows that the single upward reflection of emissions reaching the ground results in an increase of the emissions concentration at the downwind receptor z by the addition of the term g_2 to the term g_1.

It can be seen in equation (53) that the term g_1 (when there is no vertical dispersion barrier) involves the vertical distance travelled from the plume centerline height to the receptor height at z, which distance is equal to (z - H).[†] Similarly, it can be seen in equation (54) that the term g_2 (when there is a single upward reflection of emissions from the ground) involves the vertical distance travelled from the plume centerline height to the ground and back up to the receptor height at z, which distance is equal to (z + H).[†]

In the case of a trapped plume, where an inversion aloft constitutes a barrier to upward vertical dispersion, the emissions reaching the base of the inversion are reflected downwards. Thus, the emissions become "trapped" between the ground barrier below and the inversion barrier above and are subjected to an infinite number of reflections. In effect, the emissions are "bounced" back and forth between the ground and the base of the inversion aloft. Each "bounce" or reflection adds another term to $g_1 + g_2$ and each new term involves a travel distance. Thus, we can have:

$$(55) \quad C = \frac{Q}{u} \cdot \frac{f}{\sigma_y \sqrt{2\pi}} \cdot \frac{g_1 + g_2 + g_3}{\sigma_z \sqrt{2\pi}}$$

where: g_3 = an exponential term which accounts for the multiple reflections of a trapped plume

The analytical expression of the exponential term g_3 is developed in the next section.

MULTIPLE REFLECTIONS

The concept of an imaginary plume located beneath the ground surface (see Figure 10) was employed in Chapter 2 to develop the reflection dispersion term of:

$$g_2 = \exp[-(z + H)^2/(2\sigma_z^2)]$$

to account for the single upward reflection of emissions from the ground. The distance from the ground to the centerline of the imaginary plume was taken to be the same as from the ground to the centerline of the actual plume.

For the case of a trapped plume, Figure 36 presents the concept of multiple reflections from the first few of a series of imaginary plumes beneath the ground and above the inversion lid (i.e., the base of the inversion aloft):

- Emissions dispersing directly downward to the receptor from the centerline of the actual (i.e., real) plume 1 would arrive at the point a at elevation z after travelling a distance of (H - z).

- The centerline of the first below-ground imaginary plume 2 is at the same distance H

[†] (z - H) and (H - z) are numerically equivalent. Similarly, (z + H) and (H + z) are also numerically equivalent.

Chapter 8: Trapped Plumes

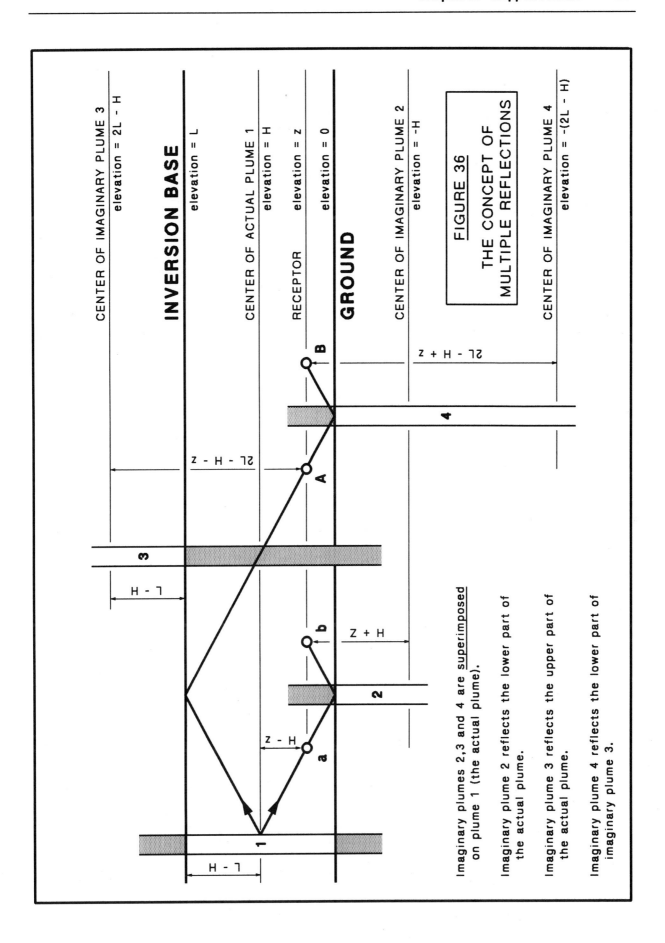

FIGURE 36
THE CONCEPT OF MULTIPLE REFLECTIONS

Imaginary plumes 2, 3 and 4 are superimposed on plume 1 (the actual plume).

Imaginary plume 2 reflects the lower part of the actual plume.

Imaginary plume 3 reflects the upper part of the actual plume.

Imaginary plume 4 reflects the lower part of imaginary plume 3.

from the ground as is the actual plume centerline. Emissions dispersing upward to the receptor from the centerline of the imaginary plume 2 would arrive at the point b at elevation z after travelling a distance of (H + z).

- Emissions dispersing upward from the actual plume cannot penetrate the inversion base and are reflected downward to the point A at elevation z.

- The centerline of the imaginary plume 3, which is within the inversion, is at the same distance (L - H) from the inversion lid as is the actual plume centerline. Emissions dispersing downward from the centerline of the imaginary plume 3 would arrive at the point A at elevation z after travelling a distance of (2L - H - z).

- The centerline of the second below-ground imaginary plume 4 is at the same distance (2L - H) below the ground as the centerline of the imaginary plume 3 is above the ground. Emissions dispersing upward to the receptor from the centerline of the imaginary plume 4 would arrive at point B at elevation z after travelling a distance of (2L - H + z).

- The total vertical distance travelled to arrive at each of the successive points a, b, A and B is the same as the vertical distance between the given point and the centerline of the previous plume. Those respective centerline-to-receptor vertical distances can be used to determine the additive g terms for use in equation (55).

The actual plume and all of the imaginary plumes are considered as being superimposed one upon the other in the same physical location, and points a, b, A and B are considered as being a single point located at the same downwind distance from the source stack. The various plumes and points have been "pulled apart" in Figure 36 only for the purpose of making it easier to visualize the multiple reflections or bounces that determine the total distance that emissions from a trapped plume must travel from the centerline of the plume to the receptor.

Figure 36 depicts only the very first few of the infinite series of reflections involved in a trapped plume. None-the-less, it serves to illustrate the point that the vertical distances travelled by the plume emissions from the actual plume centerline to the downwind receptor, for each of the individual reflections in the infinite series, determine the exponential g terms in equation (55).

Our problem now is to expand the concept illustrated in Figure 36 to an infinite series and to obtain a _generalized_ analytical expression for the travel distance required by each reflection in the series. Summing the g terms involving those distances will then provide an overall dispersion term for equation (55) which will account for the multiple reflections of a trapped plume.

First, we will differentiate between those emissions which are originally dispersing _downward_ from the actual real plume and those which are originally dispersing _upward_ from the actual real plume. The path of the emissions originally travelling downward from the actual real plume centerline will be traced by the "pulled out" series of receptor points a, b, c, d, ... in the sketch on the next page. The path of the emissions originally travelling upward from the actual real plume centerline will be traced by the "pulled out" series of receptor points A, B, C, D, ... in the same sketch. The vertical distances travelled from the starting point

Chapter 8: Trapped Plumes

(at the centerline of the actual real plume) to each of the respective, superimposed receptor points are listed immediately below the sketch.

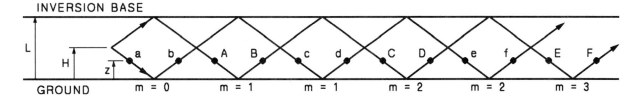

a = (H - z) = (H - z)		A = (L - H) + (L - z) = 2L - (z + H)		
b = a + 2z = (H + z)		B = A + 2z		= 2L + (z - H)
c = a + 2L = (H - z) + 2L		C = A + 2L		= 4L - (z + H)
d = b + 2L = (H + z) + 2L		D = B + 2L		= 4L + (z - H)
e = c + 2L = (H - z) + 4L		E = C + 2L		= 6L - (z + H)
f = d + 2L = (H + z) + 4L		F = D + 2L		= 6L + (z + H)
•		•		
•		•		
•		•		
r = (H - z) + 2mL		R = 2mL - (z + H)		
s = (H + z) + 2mL		S = 2mL + (z - H)		

where: m = 0 to ∞   where: m 1 to ∞

The <u>arbitrary</u> index m shown in the above sketch, when defined as below, satisfies the generalized equations given just above for any superimposed receptor points r, s, R and S:

m = 0 is used for the travel distances to before and after the first ground reflection

m = 1 is used for the travel distances to before and after each of the next two ground reflections

m = 2 is used for the travel distances to before and after each of the next two ground reflections

etc.

For example, for point F indexed by m = 3, and for point a indexed by m = 0 :

F = 2mL + (z - H) = 6L + (z - H)   which agrees with the above array
a = (H - z) + 2mL = (H - z)   which also agrees with the above array

Since each of the additive exponential g terms being sought for equation (55) involves the square of the travel distances, we can now write:

$$g = \sum_{m=0}^{m=\infty} \{\exp[-(H - z + 2mL)^2/(2\sigma_z^2)] + \exp[-(H + z + 2mL)^2/(2\sigma_z^2)]\}$$

$$+ \sum_{m=1}^{m=\infty} \{\exp[-(2mL - z - H)^2/(2\sigma_z^2)] + \exp[-(2mL + z - H)^2/(2\sigma_z^2)]\}$$

Rearranging, reversing some of the signs[†] and separating out the first two values of the first summation (for m = 0), leaving us with four terms to be summed from m = 1 to m = ∞:

$$g = \exp[-(z-H)^2/(2\sigma_z^2)] + \exp[-(z+H)^2/(2\sigma_z^2)]$$

$$+ \sum_{m=1}^{m=\infty} \{\exp[-(z-H-2mL)^2/(2\sigma_z^2)] + \exp[-(z+H+2mL)^2/(2\sigma_z^2)]$$

$$+ \exp[-(z+H-2mL)^2/(2\sigma_z^2)] + \exp[-(z-H+2mL)^2/(2\sigma_z^2)]\}$$

We have now derived the term g_3 in equation (55) and have a **GENERALIZED DISPERSION EQUATION FOR A CONTINUOUS POINT-SOURCE PLUME (EITHER TRAPPED OR NON-TRAPPED):**

(56) $$C = \frac{Q}{u} \cdot \frac{f}{\sigma_y \sqrt{2\pi}} \cdot \frac{g_1 + g_2 + g_3}{\sigma_z \sqrt{2\pi}}$$

where: f = crosswind dispersion parameter
$= \exp[-y^2/(2\sigma_z^2)]$

g = vertical dispersion parameter
$= g_1 + g_2 + g_3$

$g_1 = \exp[-(z-H)^2/(2\sigma_z^2)]$

$g_2 = \exp[-(z+H)^2/(2\sigma_z^2)]$

$$g_3 = \sum_{m=1}^{m=\infty} \{\exp[-(z-H-2mL)^2/(2\sigma_z^2)]$$

$$+ \exp[-(z+H+2mL)^2/(2\sigma_z^2)]$$

$$+ \exp[-(z+H-2mL)^2/(2\sigma_z^2)]$$

$$+ \exp[-(z-H+2mL)^2/(2\sigma_z^2)]\}$$

When the height of the inversion lid L is very large, the term g_3 becomes essentially zero, the vertical dispersion parameter g becomes equal to $g_1 + g_2$, and equation (56) is then equivalent to the generalized Gaussian dispersion equation (13) derived for a non-trapped plume in Chapter 2. **The important point regarding equation (56) is that it applies to either trapped or non-trapped plumes ... whereas equation (13) applies only to non-trapped plumes.**

The sum of the exponential terms in g_3 converges to a final value quite rapidly. For most cases, the summation of the series with m = 1, m = 2, and m = 3 provides an adequate evaluation of the series.

The multiple reflection equation (56) often appears in the literature, with each author

---

[†] $(z - H - 2mL)^2$ is equivalent to $(H - z + 2ml)^2$, and $(z + H - 2mL)^2$ is equivalent to $(2mL - z - H)^2$.

presenting it in different forms using different symbols and nomenclature. Equation (56) as derived in this chapter is completely equivalent to the expression originated by Hewson and his colleagues[10,54] with the one exception that their expression used the Sutton dispersion coefficients (see Chapter 3) rather than Pasquill's Gaussian dispersion coefficients.

Shum et al[20] derived a special form of equation (56) for ground-level concentrations ($z = 0$) when the vertical dispersion parameter is evaluated only for $m = 0$ and $m = 1$. Using the same symbols as for equation (56), their special form includes:

$$g_1 + g_2 = 2 \exp[-H^2/(2\sigma_z^2)]$$

$$g_3 = 2\{\exp[-(-H + 2L)^2/(2\sigma_z^2)] + \exp[-(-H - 2L)^2/(2\sigma_z^2)]\}$$

The above special form equations of Shum et al are readily obtained from equation (56) by using $z = 0$, and by <u>assuming</u> that evaluation of the vertical dispersion parameter g_3 using only $m = 0$ and $m = 1$ is adequate. As shown in Figure 37 (discussed in the next section of this chapter), the use of $m = 0$ and $m = 1$ is adequate only for those receptor distances at which σ_z/L is 1.0 or less.

Slade's equation 3.134[7] is the same as the special form of Shum et al, and Slade describes it as such. However, Slade does not explain that it is based on $z = 0$ and evaluation only through $m = 1$. Slade's equation 3.134 also excludes the crosswind dispersion factor and is therefore applicable only for ground-level concentrations at the plume centerline.

The EPA's HIWAY dispersion model[16] includes a multiple reflection equation expressed in the identical form of equation (56). However, the presentation of the equation in the HIWAY user's guide has an error in the sign of one of the terms in the g_3 summation.

Novak and Turner[55] present the multiple reflection equation which is used in the EPA's RAM dispersion model. Expressed with the symbols used in this chapter, their equation is:

$$g_1 + g_2 + g_3 = \sum_{m=-\infty}^{m=\infty} \{\exp[-(z + H + 2mL)^2/(2\sigma_z^2)] + \exp[-(z - H + 2mL)^2/(2\sigma_z^2)]\}$$

The above expression is completely equivalent to equation (56), and Novak and Turner recommend evaluation through $m = 4$.

Yamartino[56] presents the multiple reflection equation in yet another method which is completely equivalent to equation (56):

$$g_1 + g_2 + g_3 = \sum_{m=-\infty}^{m=\infty} \exp[-(z \pm H + 2mL)^2/(2\sigma_z^2)]$$

The $\pm$ notation in Yamartino's equation means that the summation of the multiple reflections includes the exponential terms evaluated for the positive value of H and for the negative value of H ... which makes Yamartino's equation exactly the same as Novak and Turner's equation.

WHERE VERTICAL DISPERSION OF TRAPPED PLUMES BECOMES UNIFORM

The vertical dispersion parameter $g = g_1 + g_2 + g_3$ is defined by equation (56). Figure 37 displays the relationship between the vertical dispersion parameter for trapped plumes and the ratio σ_z/L as a function of the ratio H_e/L. The ratio σ_z/L is a measure of the downwind distance from the source as a fraction of the inversion lid height (i.e., the vertical mixing depth). Similarly, the ratio H_e/L is a measure of the plume centerline height as a fraction of the mixing depth.

Figure 37 is based on the vertical dispersion parameter defined by equation (56) and for ground-level receptors. Study of Figure 37 reveals that when the ground-level receptor downwind distance is at $\sigma_z/L = 1.2$ or more:

- the vertical dispersion parameter of trapped plumes is independent of the plume centerline height (H_e), and

- the vertical dispersion parameter divided by σ_z/L has a constant value of 2.506.

Thus, $\qquad gL/\sigma_z = 2.506 \qquad$ for $\sigma_z/L \geq 1.2$

And, $\qquad \dfrac{g}{\sigma_z \sqrt{2\pi}} = \dfrac{2.506}{L\sqrt{2\pi}} = \dfrac{1}{L}$

Therefore, for values of $\sigma_z/L \geq 1.2$, equation (56) simplifies to:

$$(57) \qquad C = \frac{Q}{u} \cdot \frac{f}{\sigma_y \sqrt{2\pi}} \cdot \frac{1}{L}$$

Equation (57) is completely equivalent to equation (21) presented in the brief discussion of trapped plumes in Chapter 2. In effect, at a downwind distance of $\sigma_z/L = 1.2$, a trapped plume becomes uniformly dispersed in the vertical dimension and the only factor affecting plume concentrations beyond that point is crosswind dispersion.

The downwind distance (in km) at which σ_z/L becomes 1.2, for various heights L of the mixing lid, is presented in Table 16 as a function of the rural and urban σ_z values for the various Pasquill stability classes. As shown in Table 16, the distance at which the vertical dispersion of trapped plumes in rural terrain becomes uniform ranges from 1 to 17 km for stability classes A and B with inversion lid heights up to 2000 m, and from 12 to 26 km for stability class C with inversion lid heights up to 1000 m. For classes D and E, the point of uniform vertical dispersion occurs at distances far beyond where the application of the Gaussian dispersion equations can be justified.

Novak and Turner[55], discussing the EPA's RAM dispersion model, recommend the point of uniform vertical dispersion for trapped plumes be taken at $\sigma_z/L = 1.6$. Hoffnagle and Bass[57] indicate that the uniform vertical dispersion of trapped plumes occurs at $\sigma_z/L = 0.8$. Numerically, the difference between defining the point of uniform vertical dispersion as occurring at a σ_z/L of 0.8, 1.2 or 1.6 is not significant. However, Figure 37 makes it quite clear that 1.2 is the correct choice.

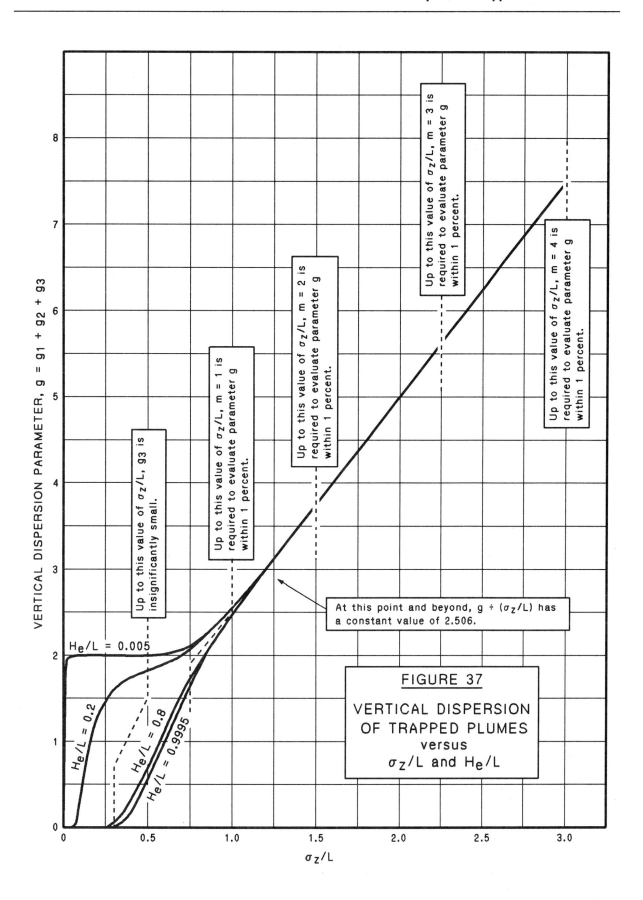

FIGURE 37

VERTICAL DISPERSION OF TRAPPED PLUMES versus σ_z/L and H_e/L

Figure 37 also illustrates how the vertical dispersion of a plume is affected by the multiple reflections term g_3 at those distances where the plume is not yet uniformly dispersed (i.e, where σ_z/L is less than 1.2):

- For plumes well below the mixing lid (i.e., H_e/L of 0.2 to 0.3 or less), the multiple reflections term g_3 does not become significant until the receptor downwind distance is at σ_z/L of 0.5 or more.

- For plumes which are just below the mixing lid (i.e., H_e/L of 0.7 to 0.8 or more), the multiple reflections term g_3 becomes significant at receptor downwind distances for which σ_z/L is 0.3 or more.

In other words, the closer the plume centerline approaches the mixing lid above the plume, the more pronounced is the effect of plume trapping upon vertical dispersion and the shorter is the downwind distance at which that effect becomes significant. Obviously, the word "significant" is a subjective term. It is defined in Figure 37 as being the point at which g_3 is at least 1 percent of the overall vertical dispersion parameter g.

TABLE 16:

DOWNWIND DISTANCE (in km) AT WHICH σ_z/L BECOMES 1.2

	L = 500 m	L = 1000 m	L = 2000 m
Rural σ_z values:			
Stability A	1.15	1.60	2.20
Stability B	4.80	9.00	16.80
Stability C	12.30	26.00	56.00
Stability D	> 100.00	> 100.00	> 100.00
Urban σ_z values:			
Stability A-B	1.55	2.65	4.30
Stability C	3.00	6.00	12.00
Stability D	7.90	25.00	92.00
Stability E-F	86.00	> 100.00	> 100.00

TURNER'S APPROXIMATION

As discussed in Chapter 2, Turner[9] proposed a method for predicting the dispersion of a trapped plume which has become rather widely accepted:

- The impingement of a trapped plume upon the base of an inversion begins to significantly affect the plume dispersion at a downwind distance x_L where:

    (20) $\quad L = H_e + 2.15\, \sigma_z$

- The vertical dispersion of a trapped plume is assumed to become uniform at a downwind distance of 2 x_L.

- At downwind distances between x_L and $2\,x_L$, interpolation of the trapped plume dispersion can be made from a log-log plot of the concentrations at points x_L and $2\,x_L$. Thus, at $x < x_L$:

(54) $$C = \frac{Q}{u} \cdot \frac{f}{\sigma_y \sqrt{2\pi}} \cdot \frac{g_1 + g_2}{\sigma_z \sqrt{2\pi}}$$

And, at $x > 2\,x_L$:

(57) $$C = \frac{Q}{u} \cdot \frac{f}{\sigma_y \sqrt{2\pi}} \cdot \frac{1}{L}$$

At $x_L \leq x \leq 2\,x_L$, C is logarithmically interpolated between the value at x_L and the value at $2\,x_L$.

- Rearranging Turner's definition of where x_L occurs, we obtain:

(20a) $\quad 1 - (H_e/L) = 2.15\,\sigma_z/L$

In effect, equation (20a) defines the point at which plume trapping becomes significant for various values of H_e/L using Turner's approximation and accepting his definition of significant. As discussed earlier in this section, Figure (37) defines the points at which plume trapping becomes significant using the more rigorous multiple reflections equation (56) for the dispersion of a trapped or non-trapped plume. In terms of defining the point at which plume trapping becomes significant, equation (56) and Turner's approximation are compared below:

H_e/L	Using Turner's equation (20a)	Using more rigorous equation (56)
	The point at which plume trapping becomes significant occurs at the distance where:	
0.0	$\sigma_z = 0.47\,L$	$\sigma_z = 0.50\,L$
0.2	$\sigma_z = 0.37\,L$	$\sigma_z = 0.50\,L$
0.8	$\sigma_z = 0.09\,L$	$\sigma_z = 0.30\,L$

As seen in this comparison, when Turner's equation (20a) is used, the point at which plume trapping becomes significant occurs where $\sigma_z = 0.47\,L$ <u>only</u> if the plume's centerline is at ground-level (i.e., $H_e/L = 0$). For that one specific situation, Turner's approximation and equation (56) are in good agreement. However, for any elevated plume centerlines, Turner's approximation deviates quite considerably from equation (56).

Now that most dispersion calculations are performed by computers, there is no real need for using any approximation. Nor is there any need to define x_L and to designate where the uniform dispersion equation (57) for a trapped plume can be used. It is much simpler to just use the generalized dispersion equation (56) for trapped or non-trapped plumes. All that is required is the selection of the proper value for the mixing lid height L. In those cases where an inversion aloft is not present, or where one wishes to exclude plume trapping, an L value of 20,000 meters will make the multiple reflections dispersion term g_3 essentially zero ... and equation (56) then becomes identical with (54) for non-trapped plumes.

GROUND-LEVEL CONCENTRATIONS FROM TRAPPED PLUMES

If used only to calculate ground-level concentrations, equation (56) can be simplified by letting z = 0:

$$g_1 = \exp[-(-H)^2/(2\sigma_z^2)]$$

$$g_2 = \exp[-(H)^2/(2\sigma_z^2)]$$

$$g_3 = \sum_{m=1}^{m=\infty} \{\exp[-(-H - 2mL)^2/(2\sigma_z^2)] + \exp[-(H + 2mL)^2/(2\sigma_z^2)] + \exp[-(-H + 2mL)^2/(2\sigma_z^2)] + \exp[-(H - 2mL)^2/(2\sigma_z^2)]\}$$

And, therefore, we have these simplified, component terms within the overall vertical dispersion parameter g:

$$g_1 + g_2 = 2\exp[-H^2/(2\sigma_z^2)]$$

$$g_3 = 2\sum_{m=1}^{m=\infty}\{\exp[-(H \pm 2mL)^2/(2\sigma_z^2)]\}$$

By substituting the simplified terms into equation (56), we obtain:

(a) $\quad C = \dfrac{Q}{u\,\sigma_y\,\sigma_z\,\pi}\{\exp[-y^2/(2\sigma_y^2)]\}\{\exp[-H^2/(2\sigma_z^2)] + \sum_{m=1}^{m=\infty}\exp[-(H \pm 2mL)^2/(2\sigma_z^2)]\}$

By referring to Figure 37, equation (a) can be further simplified:

When $\sigma_z/L < 1.2$, evaluating the vertical dispersion parameter through m = 2 is sufficient:

(b) $\quad C = \dfrac{Q}{u\,\sigma_y\,\sigma_z\,\pi}\{\exp[-y^2/(2\sigma_y^2)]\}\{\exp[-H^2/(2\sigma_z^2)] + \sum_{m=1}^{m=2}\exp[-(H \pm 2mL)^2/(2\sigma_z^2)]\}$

When $\sigma_z/L \geq 1.2$, the vertical dispersion of trapped plumes then becomes uniform and equation (57) is applicable. Equation (57) can be rearranged thus:

$$C = \dfrac{Q}{u\,\sigma_y}\{\exp[-y^2/(2\sigma_y^2)]\}\dfrac{1}{L\sqrt{2}\sqrt{\pi}} \quad \text{where:} \quad \dfrac{1}{L\sqrt{2}\sqrt{\pi}} \cdot \dfrac{\sigma_z\sqrt{\pi}}{\sigma_z\sqrt{\pi}} = \dfrac{1.253\,\sigma_z/L}{\sigma_z\,\pi}$$

And so we can write:

(c) $\quad C = \dfrac{Q}{u\,\sigma_y\,\sigma_z\,\pi}\{\exp[-y^2/(2\sigma_y^2)]\}(1.253\,\sigma_z/L)$

Equations (b) and (c) can be combined and re-stated as shown below to obtain a **GENERALIZED DISPERSION EQUATION FOR GROUND-LEVEL CONCENTRATIONS FROM A CONTINUOUS POINT-SOURCE PLUME (EITHER TRAPPED OR NON-TRAPPED)**:

(58) $$C = \frac{Q}{u \sigma_y \sigma_z \pi} \cdot f \cdot g^*$$

where: f = crosswind dispersion parameter
   $= \exp[-y^2/(2\sigma_y^2)]$

g^* = vertical dispersion factor

$$= \exp[-H^2/(2\sigma_z^2)] + \sum_{m=1}^{m=2} \exp[-(H \pm 2mL)^2/(2\sigma_z^2)] \quad \text{for } \sigma_z/L < 1.2$$

$$= 1.253 \, \sigma_z/L \quad \text{for } \sigma_z/L \geq 1.2$$

To avoid any undue confusion, it should be noted that a comparison of equations (56) and (58) reveals that the vertical dispersion <u>parameter</u> g is equal to twice the vertical dispersion <u>factor</u> g^*. This is explained by:

Equation (56) applies for any receptor elevation, whereas equation (58) applies <u>only</u> for ground-level receptors (at $z = 0$) and ground-level concentrations. As shown on the previous page, simplification of equation (56), to calculate ground-level concentrations, resulted in extracting a factor of 2 from the component g terms. In other words:

$$C = \frac{Q}{u \sigma_y \sigma_z 2\pi} \cdot f \cdot (g = g_1 + g_2 + g_3)$$

$$= \frac{Q}{u \sigma_y \sigma_z \pi} \cdot f \cdot (g^* = g_1^* + g_2^* + g_3^*)$$

For the same reason, when $\sigma_z/L \geq 1.2$, we have $g = 2.506 \, \sigma_z/L$ and $g^* = 1.253 \, \sigma_z/L$.

As explained earlier in this section, the $\pm$ notation in many of the foregoing equations, means that the summation of the multiple reflections includes the exponential terms evaluated for the positive value of H and for the negative value of H. In other words:

$$\sum_{m=1}^{m=\infty} \exp[-(H \pm 2mL)^2/(2\sigma_z^2)] = \sum_{m=1}^{m=\infty} \{\exp[-(H + 2mL)^2/(2\sigma_z^2)] + \exp[-(H - 2mL)^2/(2\sigma_z^2)]\}$$

Chapter 9

FUMIGATION

EVOLUTION OF A PLUME FUMIGATION

When a plume embedded in an inversion layer encounters rising turbulence, the plume undergoes a rapid breakup and its vertical dispersion becomes uniform from the ground to the plume top. This behavior is referred to as "fumigation" and was discussed briefly in Chapter 2. Figure 38 depicts how a plume fumigation evolves:

- The upper sketch in Figure 38 depicts a fanning plume embedded at dawn in the "radiation inversion" layer which forms at the earth's surface during the night (see Chapter 1).

- The next sketch shows the formation of a turbulent layer beneath the inversion, caused by warming of the earth's surface in the morning. The lower, turbulent layer has a negative ambient temperature gradient, meaning that the ambient temperature in the turbulent layer decreases with increasing altitude. The lower layer is unstable because the absolute value of its ambient temperature gradient (the solid line) is greater than the adiabatic lapse rate (dashed line). See Chapter 1 for further details.

- In the next sketch, the turbulent layer has risen by noon to reach the bottom of the plume. The encounter with rising turbulence causes the plume to begin fumigating (i.e., breaking up), and the lower edge of the plume is rapidly dispersed downward.

- The final sketch shows a complete fumigation, which occurs in the afternoon when the rising turbulent layer reaches the top of the plume. At that time, the vertical dispersion of the plume is uniform from the ground to the plume top.

The ΔT values shown in Figure 38 are a measure of the heat transferred from the rising turbulent layer during the evolution of a fumigation. The significance of that heat transfer is discussed later in this chapter.

GROUND-LEVEL CONCENTRATIONS DURING FUMIGATION

On the basis that fumigation results in uniform vertical dispersion from the ground to the plume top, the ground-level concentrations during complete fumigation would be:

(59) $$C = \frac{Q}{u\,\sigma_{yf}\sqrt{2\pi}} \cdot \exp[-y^2/(2\,\sigma_{yf}^2)] \cdot \frac{1}{(H_e + 2.5\,\sigma_z)}$$

where: $H_e + 2.5\,\sigma_z$ = distance from plume top to ground, m
H_e = plume centerline height, m
σ_{yf} = crosswind dispersion coefficient during fumigation, m
σ_z = vertical dispersion coefficient during inversion conditions (stability class E or F), m

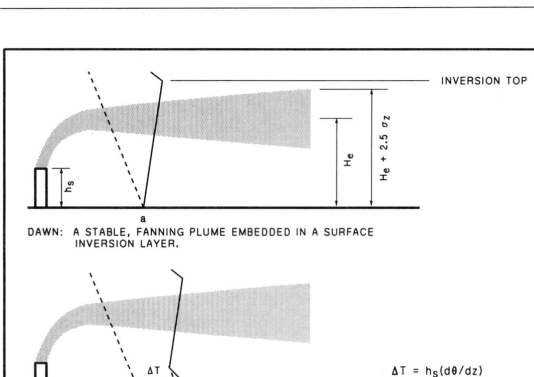

$\Delta T = h_s (d\theta/dz)$

$\Delta T = (H_e - 2.5\,\sigma_z)(d\theta/dz)$

$\Delta T = (H_e + 2.5\,\sigma_z)(d\theta/dz)$

FIGURE 38

EVOLUTION OF A PLUME FUMIGATION

Chapter 9: Fumigation

The fumigation dispersion equation (59) is quite similar to the dispersion equation (57) for a trapped plume which has travelled downwind far enough so that its vertical dispersion has become uniformly distributed within the mixing layer:

$$(57) \quad C = \frac{Q}{u\, \sigma_y \sqrt{2\pi}} \cdot \exp[-y^2/(2\sigma_y^2)] \cdot \frac{1}{L}$$

The only differences between the fumigation equation (59) and the uniform dispersion equation (57) for a trapped plume are:

- Uniform vertical dispersion occurs from the ground up to the top of the plume (i.e., $H_e + 2.5\,\sigma_z$) for a fumigated plume ... whereas it occurs from the ground up to the bottom of an inversion aloft (i.e., the mixing height L).

- The crosswind dispersion coefficient σ_{yf} for a fumigated plume includes a correction for the additional horizontal spreading caused by the intense mixing that occurs during fumigation.

As discussed in Chapter 2, the vertical width of a Gaussian plume can be expressed in terms of the dispersion coefficient σ_z. At a vertical distance of $2.15\,\sigma_z$ from the plume centerline, the plume boundary concentration is 10 percent of the centerline concentration. Thus, many dispersion modelling practitioners have used $2.15\,\sigma_z$ as the vertical half-width of a dispersing plume. However, as a matter of personal preference, Figure 38 and equation (59) both use $2.5\,\sigma_z$ to define the vertical half-width of a plume. The plume boundary concentrations are slightly more than 5 percent of the plume centerline concentration for a plume half-width of $2.5\,\sigma_z$ ... and thus the plume half-width defined in equation (59) includes more of the total plume than does a plume half-width of $2.15\,\sigma_z$.

Equation (59) is identical to Turner's equation[9] for maximum fumigation, except that Turner uses a plume half-width of only $2\,\sigma_z$ as his own preference.

CROSSWIND DISPERSION DURING FUMIGATION

Bierly and Hewson[54] suggested an approximation for obtaining the ground-level value of σ_{yf}, from which Turner[9] developed the following:

- The crosswind dispersion coefficient at the plume centerline H_e is σ_y for stability classes E or F (conditions within the inversion layer) at the receptor downwind distance x_f where complete fumigation occurs.

- The <u>crosswind</u> plume half-width is $2.15\,\sigma_y$ at the plume centerline height H_e and increases by $(H_e)(\tan 15°)$ as the plume spreads to the ground during complete fumigation.

- Thus, Turner obtained:

$$2.15\,\sigma_{yf} = 2.15\,\sigma_y + H_e(\tan 15°)$$
$$\sigma_{yf} = \sigma_y + 0.125\,H_e$$

Following the same procedure as Turner, but using 2.5 σ_y to define the plume half-width, we obtain:

(60) $\qquad \sigma_{yf} = \sigma_y + 0.11\ H_e$

where: σ_y = crosswind dispersion coefficient in meters at the plume centerline, at downwind distance x_f in meters, for the inversion conditions (stability class E or F)

σ_{yf} = ground-level, crosswind dispersion coefficient in meters, at the downwind distance x_f in meters, during fumigation

x_f = the downwind distance in meters at which the maximum fumigation occurs

H_e = height in meters of the plume centerline (i.e., the effective stack height, or the emissions height)

DOWNWIND DISTANCE TO MAXIMUM FUMIGATION

Fumigation starts when a rising turbulent layer first reaches the bottom of a fanning plume embedded in an inversion layer (see Figure 38). Maximum fumigation occurs when the rising turbulence later reaches the plume top and fumigates the entire plume. When fumigation starts at the plume bottom, the plume top is still moving downwind in the inversion layer at the windspeed u. During the time t, while the turbulence rises from the plume bottom to the plume top, the plume top travels a downwind distance x = ut from the point at which the rising turbulence first contacts the plume bottom. Similarly, during the time period t_f, while the turbulence rises from the source stack exit to the plume top, the plume top travels a distance of $x_f = ut_f$ downwind from the source stack exit.

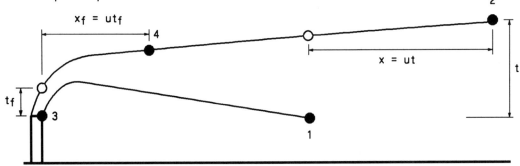

Fumigation may start at some point 1 (lower than the stack height) and result in maximum fumigation at point 2. However, the <u>minimum</u> downwind distance from the stack to the point of maximum fumigation occurs at point 4 for a fumigation starting at point 3 just at the stack exit. Thus, the minimum distance from the source stack to maximum fumigation can be defined as $x_f = ut_f$, where t_f is the time required for the turbulent layer to rise to the top of the plume at x_f.

The breakup of the inversion layer can be defined as occurring when the heat transferred from the warm, rising turbulence is enough to change the temperature gradient dT/dz of the inversion to that of a dry adiabatic lapse rate (denoted as Γ). As shown by points a and b in Figure 38, when the rising turbulence has eliminated the inversion from the ground up to the source stack height h_s, the heat transferred by that time must have been at least sufficient to create a ΔT at ground level of $h_s(d\theta/dz)$ ... since, by definition, the potential temperature gradient $d\theta/dz$ is the difference between the ambient temperature gradient dT/dZ and the adiabatic lapse rate Γ (i.e., $d\theta/dz = (dT/dz) - \Gamma$). As also shown by points a and d in Figure 38, when the rising turbulence has eliminated the inversion from the ground up to the plume top, the heat transferred by that time must have been at least sufficient to create a ΔT at ground level of $(H_e + 2.5\, \sigma_z)(d\theta/dz)$. Thus, the heat which has been absorbed per cubic meter of the inversion layer at time 0 (when the turbulence reaches the stack height) and at time t_f (when the turbulence reaches the plume top) can be expressed as:

$$q = \rho c_p \Delta T = \rho c_p (d\theta/dz)(h_s) \qquad \text{at time 0}$$

$$q = \rho c_p \Delta T = \rho c_p (d\theta/dz)(H_e + 2.5\, \sigma_z) \qquad \text{at time } t_f$$

which can be generalized as:

(61) $$q = \rho c_p \Delta T = \rho c_p (d\theta/dz)(h) = k(h)$$

  where: q = heat transferred during elimination of inversion from the ground up to height h, cal/m^3

     c_p = heat capacity of ambient air, cal/g/°K

     ρ = ambient air density, g/m^3

     $d\theta/dz$ = the potential temperature gradient within the inversion layer, °K/m

     h = the height above ground to which the inversion has been eliminated, m

     $k = \rho c_p(d\theta/dz)$ in cal/m^4

By integrating equation (61), we can determine the heat transfer required to eliminate any part of the inversion:

$$Q = \int_{h_1}^{h_2} k(h)\, dh = (k/2)(h_2^2 - h_1^2)$$

  where: Q = the heat transferred during elimination of the inversion between heights h_2 and h_1, in cal/m^2 of the inversion layer surface

Thus, the heat transferred between the stack exit height and the plume top:

(62) $$Q = 0.5\, \rho c_p (d\theta/dz)[(H_e + 2.5\, \sigma_z)^2 - h_s^2]$$

Chapter 9: Fumigation

The available heat flux (i.e., heat flow rate per square meter of horizontal surface) from the rising turbulence, derived from solar radiation, is conventionally denoted as H_0. Dividing the heat absorbed per square meter of surface of the eliminated portion of the inversion by the available heat flux, we obtain the time required to eliminate that portion of the inversion layer:

(63) $$t_f = Q/H_0 = 0.5 \, \rho c_p (d\theta/dz)[(H_e + 2.5\,\sigma_z)^2 - h_s^2]/H_0$$

where: H_0 = available heat flux, $cal/m^2/sec$

Equation (63) is identical with Turner's equation 5.5,[9] which he references to Pooler,[58] except that Turner uses 2 σ_z for the vertical plume half-width whereas equation (63) uses 2.5 σ_z.

Using ρ of 1200 g/m^3 and c_p of 0.24 $cal/g/°K$ as representative of ambient air at 20 °C, equation (63) becomes:

(64) $$t_f = 144 \, (d\theta/dz)[(H_e + 2.5\,\sigma_z)^2 - h_s^2]/H_0$$

where: t_f = time period to eliminate inversion between source stack exit and plume top, sec

$d\theta/dz$ = potential temperature gradient within the inversion (for stability class E or F), °K/m

H_e = plume centerline height at distance x_f, m

σ_z = vertical dispersion coefficient for the inversion (stability class E or F) at distance x_f, m

h_s = source stack height, m

H_0 = heat flux in the rising turbulent layer (stability class A, B or C), $cal/m^2/sec$

And the minimum downwind distance to the point of maximum fumigation is x_f, where:

(65) $$x_f = u t_f$$

where: u = wind velocity at the plume centerline height during the inversion, m/sec

Some typical values of the potential temperature gradient for use with equations (64) are:

Inversion Layer Stability Class	$d\theta/dz$ in °K/m Range	Typical
E	0.005-0.025	0.015
F	> 0.025	0.035

Equation (64) requires values of the heat flux H_0 for the rising turbulent layer beneath the inversion. As a frame of reference, Table 17 lists the solar and sky radiation received at

TABLE 17: SOLAR AND SKY RADIATION FLUXES IN THE UNITED STATES

Location	Latitude		Spring	Summer	Fall	Winter
Friday Harbor, Washington	48	(a)	4100	5715	2285	1000
		(b)	6.5	6.5	3.4	2.0
		(c)	626	881	681	500
Great Falls, Montana	47	(a)	4570	5895	2930	1640
		(b)	7.8	11.1	5.7	4.5
		(c)	589	533	438	365
Bismark, North Dakota	47	(a)	4605	5510	2645	1685
		(b)	8.3	10.0	6.2	5.4
		(c)	553	551	424	314
Minneapolis, Minnesota	45	(a)	4070	5670	2160	1530
		(b)	7.2	8.8	4.4	4.0
		(c)	568	641	489	383
Boston, Massachusetts	42	(a)	3510	4380	2225	1480
		(b)	7.5	9.8	7.0	4.8
		(c)	465	446	318	311
Salt Lake City, Utah	41	(a)	4680	5790	3060	1810
		(b)	9.2	11.9	9.2	4.4
		(c)	511	488	332	410
Indianapolis, Indiana	40	(a)	4325	5570	2785	1535
		(b)	7.4	11.4	7.0	4.3
		(c)	588	488	397	364
Columbia, Missouri	39	(a)	4085	5585	3170	1955
		(b)	9.5	11.3	7.2	5.5
		(c)	431	493	437	356
Washington, D.C.	39	(a)	4020	4920	2880	1860
		(b)	7.8	10.6	6.6	5.3
		(c)	513	465	434	349
Dodge City, Kansas	38	(a)	4915	6475	3710	2935
		(b)	8.6	11.9	9.5	7.3
		(c)	569	543	392	402
Albuquerque, New Mexico	35	(a)	5585	6425	4145	2700
		(b)	10.5	11.9	9.8	7.9
		(c)	533	542	424	343
Riverside, California	34	(a)	4775	5675	3750	2595
		(b)	8.1	10.5	9.1	8.0
		(c)	588	542	410	325
Atlanta, Georgia	34	(a)	5260	6060	3720	2580
		(b)	8.0	10.8	7.7	5.3
		(c)	661	559	481	483
New Orleans, Louisiana	30	(a)	4150	4480	3485	2305
		(b)	8.4	9.5	8.0	6.3
		(c)	497	471	437	365
San Antonio, Texas	29	(a)	4930	6250	4220	3020
		(b)	8.3	11.1	9.3	6.8
		(c)	594	563	454	447
Miami, Florida	26	(a)	4615	4760	3825	3185
		(b)	9.1	9.2	8.2	8.7
		(c)	506	516	468	366
AVERAGES OF (c) VALUES: kcal/m^2/hr of sunlight			550	545	439	380
cal/m^2/sec of sunlight			153	151	122	106

(a) values are for total solar and sky radiation received, kcal/m^2/day
(b) values are sunlight hours/day
(c) values are kcal/m^2/hour of sunlight

each of 16 locations across the United States in terms of kcal/m²/day. Table 17 also includes the average sunlight hours per day for each location. The radiation received per day was combined with the average sunlight hours per day to obtain the average radiation flux R in cal/m²/sec for each location. The average radiation flux obtained in Table 17 ranges from 106 to 153 cal/m²/sec for the four weather seasons. Assuming that all the incoming solar and sky radiation is converted to upflowing heat flux H_0, those averages represent an upper bound of maximum H_0 values in the United States. However, a large part of the solar and sky radiation is lost by reflection from clouds and by atmospheric scattering. Also, part of the solar and sky radiation is absorbed by the ground as well as by evaporation of surface water and plant moisture. Briggs[23] states that only 40 percent of the incoming radiation flux R is converted to upflowing turbulent heat flux H_0, which means that the average radiation fluxes of 106 to 153 in Table 17 translate to average H_0 values of 42 to 61 cal/m²/sec. Briggs[23] also gives some typical H_0 values which translate to:

Stability class A: H_0 = 69 cal/m²/sec
Stability classes B-C: H_0 = 43 cal/m²/sec

Turner[9] references Pooler[58] as suggesting an average H_0 of about 67 cal/m²/sec for use with equation (67). Pasquill[59] gives these values for H_0:

Stability class A (urban): H_0 = 62-96 cal/m²/sec
Stability class B (urban): H_0 = 72 cal/m²/sec
Stability class B (rural): H_0 = 60 cal/m²/sec
Stability class C (rural): H_0 = 36 cal/m²/sec

Sklarew and Wilson's EPRI report[3] give the values shown in Table 1 in Chapter 1:

Strong insolation: H_0 > 38 cal/m²/sec
Slight insolation: H_0 < 18 cal/m²/sec

Sklarew and Wilson's H_0 values amount to only 27 percent of their solar radiation fluxes as contrasted with Briggs' estimate of 40 percent.[23]

As a broad generalization of the above literature values, these values of H_0 are recommended for use with equation (64):

TABLE 18: RECOMMENDED H_0 VALUES

Turbulent Layer Stability Class	Terrain	H_0, cal/m²/sec
A	Rural	65
	Urban	80
B	Rural	55
	Urban	70
C	Rural	40
	Urban	55

As suggested by Pasquill[59], the amount of heat flux listed in Table 18 for urban areas is assumed to be higher than for rural areas by 15 cal/m^2/sec.

As a point of interest, Briggs suggested a method for estimating the incoming solar radiation flux R. Given a specific location, the solar elevation at a given time for that location, and the cloudiness fraction, Briggs[23] suggests that the incoming solar radiation flux may be estimated thus:

$$R = (2/3)(\sin \theta_{el})(1 - 0.8\ C)(S)$$

where: θ_{el} = the solar elevation angle, in degrees, at a given time and place
C = the cloudiness fraction
S = the solar constant of 323 cal/m^2/sec in a sunlit cloudless sky[60]

Assuming a solar elevation of 45 degrees and a cloudiness fraction of 0.2, Briggs' suggestion yields a solar radiation flux R of 128 cal/m^2/sec. If 40 percent of that is converted to upflowing heat flux, the resultant H_0 is 51 cal/m^2/sec which is quite consistent with the recommended values in Table 18.

CALCULATION EXAMPLE

The complete fumigation of a plume results in uniform vertical dispersion from the plume top to the ground. The ground-level concentration of fumigated plume components is calculated by equation (59), which requires σ_z as an input. To determine σ_z, the downwind distance from the source stack to the point of complete fumigation is needed. That distance is obtained from equations (64) and (65) which calculate t_f and x_f ... but equation (64) also requires σ_z as an input. Thus, the calculation of ground-level fumigation concentrations involves a reiterative, trial-and-error procedure:

- Assume a trial distance to the point of complete fumigation.
- Obtain a trial σ_z at the assumed trial distance.
- Using the trial σ_z, calculate the distance x_f to the point of complete fumigation.
- Reiterate the calculations until the calculated distance matches the previous trial distance.

A typical fumigation calculation is presented in Example 8. The first step in Example 8, as in all dispersion calculations, is to select and define the emission source specifics. The next step in the example is to define the meteorology specifics for:

- The inversion layer in which the stack gas plume was embedded just prior to the fumigation.
- The rising, turbulent layer at the ground surface which caused the fumigation by heating and breaking up the inversion layer.

After defining the problem specifics, the initial steps in the example are:

- Calculation of the stack exit gas flow and the pollutant SO_2 emission rate.
- Calculation of the stack gas buoyancy and stability parameters, as well as the

downwind distance to the point of maximum plume rise as defined by Briggs.[29]
- Conversion of the ground-level windspeed to the windspeed at the source stack exit height under the inversion's atmospheric stability class.

The next step is the assumption of a first trial distance to the point of complete fumigation. The plume rise, Δh, is calculated at the trial distance using the pertinent Briggs equation for the inversion's stability class (E or F). The plume rise is added to the stack height to obtain the effective stack height H_e. The ground-level windspeed is then converted to windspeed at the effective stack height under the inversion's stability class. Next, the dispersion coefficient σ_z is determined at the trial distance for the pertinent terrain (rural or urban). Finally, equations (64) and (65), as well as the windspeed at the effective stack height, are used to calculate the distance to the point of complete fumigation. If the calculated distance to the point of complete fumigation does not match the assumed trial distance, a second trial distance is assumed and the calculations are reiterated until a reasonable match is obtained.

The distance to the point of complete fumigation is typically greater than 5-10 km. At such distances, most plumes will have attained their maximum plume rise as defined by Briggs[29] and, therefore, the effective stack height generally will not change during the reiterative calculations. However, a check of the trial distance versus the downwind distance to the point of maximum plume rise (i.e., Briggs' 3.5 x*) should be made during each reiteration.

Figure 39 presents the results of several fumigation calculations for the specifics defined in Example 8. The lower part of Figure 39 plots the resultant ground-level concentrations during fumigation versus various ground-level windspeeds within the inversion layer. The upper part of Figure 39 plots the same resultant ground-level concentrations during the fumigation versus the various windspeeds at the effective stack height (i.e., the plume centerline height) within the inversion layer.

Figure 39 applies only for the specifics defined in Example 8, but it qualitatively illustrates an important point. **For a given ground-level windspeed, the more stable is the inversion in which the plume was imbedded prior to the fumigation, the <u>closer</u> is the distance to the point of complete fumigation.** A less stable inversion has a larger vertical dispersion coefficient and a higher effective stack height (because of a higher plume rise), both of which appear in equation (64) and would lead to a longer fumigation time and hence a further distance to the point of complete fumigation. A more stable inversion has a higher potential temperature gradient which also appears in equation (64) and would also lead to a longer fumigation time and to a further distance to the point of complete fumigation. Evidently, higher dispersion coefficients and higher effective stack heights have more effect in equation (64) than higher potential temperature gradients. Thus, we can generalize:

Stability class of the inversion	F	E
Relative stability	more	less
θ/dz	higher	lower
Windspeed conversion exponent	higher	lower
Windspeed at ground-level	same	same
Windspeed at plume centerline height	higher	lower
Plume centerline height	lower	higher
σ_z	smaller	larger
Distance to complete fumigation point	closer	farther
Fumigation ground-level concentrations	higher	lower

Chapter 9: Fumigation 143

EXAMPLE 8: FUMIGATION CALCULATION

Calculate the downwind distance at which maximum fumigation occurs, and the ground-level SO_2 concentration at that point.

GIVEN CONDITIONS AND PROBLEM SPECIFICS:

A point-source plume derived from the stack of a 500 MW coal-fired power plant operating at an 80 percent load factor and:

Stack exit temperature	330 °F	(166 °C)
Stack height	700 ft	(213 m)
Gross heating value (GHV) of coal	11,000 Btu/lb	
Coal sulfur content	2 wt %	
Plant thermal efficiency	34 % based on coal's GHV	
Plant heat input (i.e., fuel fired)	10,044 Btu/(KW-hr of output)	
Stack gas flow	12,000 SCF/(10^6 Btu of coal)[†]	
Terrain	rural	

Inversion conditions preceding the fumigation:

Atmospheric stability	Pasquill class F	
Ground-level windspeed	1.00 knots	(0.515 m/sec)
Potential temperature gradient, $d\theta/dz$	0.03 °K/m	
Windspeed conversion exponent, n	0.30	(see Table 13)
Ambient temperature	70 °F	(21 °C)

Surface turbulent layer rising into the inversion:

Atmospheric stability	Pasquill class A
Heat flux, H_0	65 cal/m^2/sec

CALCULATIONS:

Input coal	= (500 MW)(0.8)(1000 KW/MW)(10,044 Btu/KW-hr)	= 4,020 ×10^6 Btu/hr
	= (4,020 × 10^6 Btu/hr)/(11,000 Btu/lb)	= 365,450 lbs/hr
Input sulfur	= (2/100)(365,450 lbs/hr)	= 7,309 lbs/hr
Output SO_2	= (2 lbs SO_2/lb S)(7,309 lbs S/hr)(454/3600)	= 1,843 g/sec
Stack gas flow	= (12,000 SCF/10^6 Btu of coal)(4,020)	= 48.24 × 10^6 SCF/hr[†]
	= (48.24 × 10^6)(330 + 460)/(60 + 460)	= 73.29 × 10^6 ft^3/hr
	= (73.29 × 10^6)/(35.31 ft^3/m)(3600 sec/hr)	= 577 m^3/sec

Buoyancy factor, F
$$= (gV_s/\pi)(T_s - T_a)/T_s$$
$$= (9.807)(577/\pi)(166 - 21)/(166 + 273) \qquad = 595 \ m^4/sec^3$$

Stability factor, s
$$= (g/T_a)(d\theta/dz)$$
$$= (9.807)(0.03)/(21 + 273) \qquad = 0.001 \ sec^{-2}$$

Distance to maximum plume rise
$$= 3.5 \ x^* = 119 \ F^{0.4} \qquad (\text{for } F \geq 55 \ m^4/sec^3)$$
$$= 119(595)^{0.4} \qquad = 1,532 \ m$$

Windspeed at the stack height
$$= 0.515(213/10)^{0.30} \qquad = 1.3 \ m/sec$$

[†] SCF is a standard cubic foot of gas, measured at 60 °F and 1 atmosphere

Chapter 9: Fumigation

Parameter $1.84\ u\ s^{-1/2}$
$$= (1.84)(1.3)(0.001)^{-1/2} = 76\ m$$

Assume a trial distance to maximum fumigation of x = 50 km (which is > 3.5 x^*):

Since $1.84\ u\ s^{-1/2} < 3.5\ x^*$, and since $x > 1.84\ u\ s^{-1/2}$:

$$\Delta h = 2.4(F/us)^{1/3} = (2.4)[595/(1.3)(0.001)]^{1/3} = 185\ m$$
$$H_e = 213 + 185 = 398\ m$$

Windspeed at the effective stack height H_e
$$= 0.515(398/10)^{0.30} = 1.56\ m/sec$$

Vertical dispersion coefficient σ_z
  = 79 m at 50 km downwind, stability F and rural terrain   (from Figure 17)

Using equations (64) and (65):

$$t_f = 144(0.03)[(398 + 2.5\ \sigma_z)^2 - 213^2]/65 = 20{,}553\ sec = 5.7\ hrs$$

$$x = ut_f = 1.56(20{,}553) = 32{,}063\ m$$
  = 32.1 km which is lower than the assumed trial x of 50 km

Assume a trial distance to maximum fumigation of x = 29 km (which is > 3.5 x^*):

The effective stack height H_e remains 398 m and the windspeed at the effective stack height remains 1.56 m/sec, both the same as above for the first trial.

σ_z = 68 m at 29 km downwind, stability F and rural terrain   (from Figure 17)

$$t_f = 144(0.03)[(398 + 170)^2 - 213^2]/65 = 18{,}427\ sec = 5.1\ hrs$$

$$x = ut_f = 1.56(18{,}427) = 28{,}746\ m$$
  = 28.7 km which checks with the assumed trial x of 29 km

σ_y = 680 m at 29 km downwind, stability F and rural terrain   (from Figure 18)

Using equation (60), the crosswind dispersion coefficient during fumigation is

$$\sigma_{yf} = 680 + 0.11(398) = 724\ m$$

Using equation (59) for the uniform, vertical dispersion of a fumigated plume with y = 0 to obtain the centerline pollutant concentration:

$$C = Q/[u\ \sigma_{yf}\ (2\pi)^{1/2}\ (H_e + 2.5\ \sigma_z)]$$
$$= (1{,}843\ g/sec)(10^6\ \mu g/g)/(1.56)(724)(2.51)(398 + 170) = 1{,}145\ \mu g/m^3$$

Thus, maximum fumigation occurs 28.7 km downwind of the source stack and creates a ground-level centerline SO_2 concentration of 1,145 $\mu g/m^3$.

Chapter 9: Fumigation 145

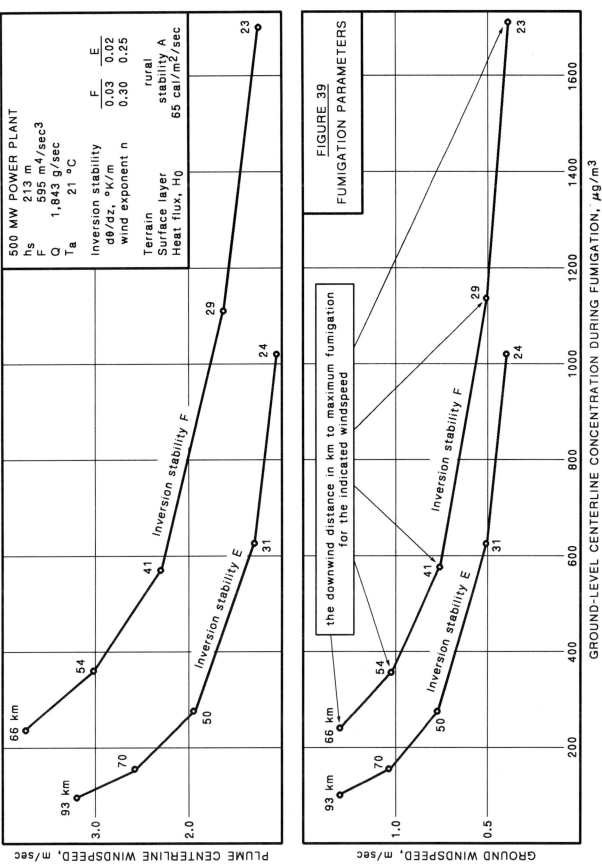

FIGURE 39
FUMIGATION PARAMETERS

Figure 39 also illustrates that the difference between the windspeed at ground-level and at the plume centerline height has a significant effect on the downwind distance to the point of maximum fumigation. The upper part of Figure 39 indicates that the windspeed at the plume centerline height (for the specific case in point) would have to be less than 1 m/sec for maximum fumigation to occur any closer to the source stack than 23 km ... as contrasted with the lower part of Figure 39 which indicates that the ground-level windspeed would have to be less than 0.4 m/sec for maximum fumigation to occur closer than 23 km.

TIME-AVERAGING OF A FUMIGATION

The TVA,[2] Turner[9] and Lyons[23] indicate that fumigations generally persist for a relatively short period only ... perhaps 30 to 45 minutes. Thus, the determination of an average ground-level concentration over an hour period, during which a maximum fumigation occurred, is quite difficult. One might time-average these combinations:

- 30 minutes of fumigation concentration plus 30 minutes of the subsequent turbulent condition concentration (for stability class A, B or C), or

- 45 minutes of fumigation concentration plus 15 minutes of the subsequent turbulent condition concentration.

Either of the above combinations poses the problem of deciding if the dispersion coefficients used to calculate the concentrations represent 1-hour or 10-minute averages or some intermediate time-average. This is the same problem discussed in Chapter 5 and at the end of Chapter 7. We can express the options as:

(66) $$C_{1-hr} = C_f k_f \left(\frac{t_f}{60}\right) + C_t k_t \left(\frac{60 - t_f}{60}\right)$$

where: C_{1-hr} = 1-hour average concentration during which a fumigation occurred
C_f = the calculated fumigation concentration, which represents some time-average period p
C_t = the calculated turbulent condition concentration, which represents some time-average p
t_f = the time average, in minutes, over which the fumigation is taken to persist
k_f, k_t = correction factors from time-average period p to time periods t_f and $(60 - t_f)$ as discussed in Chapter 5

Thus, the options in defining C_{1-hr} for a fumigation period include:

- Selection of the time-average period p represented by the Gaussian dispersion coefficients (ranging from 10-minutes to 1-hour).
- Selection of the fumigation persistence time t_f (which typically ranges from 30 to 45 minutes).
- Selection of the method for determining the time-average correction factors k_f and k_t as discussed in Chapter 5.

OTHER METHODS

Equation (64) for obtaining the time t_f required to eliminate an inversion layer was published by Pooler[58] in essentially the same form as derived herein. Two other equations have been proposed: (a) one by Hewson and his colleagues,[10] and (b) one used by the TVA.[2]

(a) Hewson's proposed equation[10] can be written as:

$$t_f = (0.25/k)(H_t^2 - H_b^2)$$

Defining the top of the inversion layer H_t as H_e plus $2.5\ \sigma_z$, and the bottom of the layer H_b as the stack height h_s, then:

$$t_f = (0.25/k)[(H_e + 2.5\ \sigma_z)^2 - h_s^2]$$

Equating this expression to equation (64), we obtain:

$$\text{Hewson's } k = H_0/(576)(d\theta/dz)$$

where: k = eddy diffusivity for heat, m^2/sec

For an inversion with $d\theta/dz$ of 0.03 broken up by rising turbulence with an H_0 of 65, we find:

$$\text{Hewson's } k = 3.8\ m^2/sec$$

which agrees well with Turner's suggested value[9] of 3.0 for Hewson's parameter k.

(b) The TVA uses an equation[2] that is quite similar to the one proposed by Hewson:

$$t_f = (0.25/K)\rho c_p(H_t^2 - H_b^2)$$

Using a plume vertical half-width of $2.5\ \sigma_z$, then:

$$t_f = (0.25/K)\rho c_p[(H_e + 2.5\ \sigma_z)^2 - h_s^2]$$

Equating this expression to equation (64) and using values for air of $\rho = 1{,}200\ g/m^3$ and $c_p = 0.24\ cal/g/°K$, we obtain:

$$\text{TVA's } K = 0.5\ H_0/(d\theta/dz)$$

where: K = eddy conductivity, $[(cal/m)/(sec)]/°K$

For an inversion with $d\theta/dz$ of 0.03 broken up by rising turbulence with an H_0 of 65, we find:

$$\text{TVA's } K = 1{,}083\ [(cal/m)/(sec)]/°K$$

and for an inversion with $d\theta/dz$ of 0.01 broken up by rising turbulence with an H_0 of 65:

$$\text{TVA's } K = 3{,}250\ [(cal/m)/(sec)]/°K$$

The TVA's values of K range from 1,200 to 8,000 (for $d\theta/dz$ ranging from 0.03 to 0.01), which is fairly consistent with values obtained above.

As can be seen in the foregoing discussion, Hewson's equation[10] and the TVA's equation[2] are essentially the same as equation (64) derived from Pooler's publication[58] when the values of H_0, k and K are consistent with one another. In using the three equations to obtain comparative fumigation calculation results, it is also most important to be consistent in:

- Calculating the plume rise to determine the effective stack height H_e
- Categorizing the atmospheric stability classes of the inversion and the turbulent layers
- Obtaining the dispersion coefficients

Otherwise, as evidenced by Gutfreund's study,[61] the results can be highly disparate.

For the most part, this chapter is devoted to fumigations which occur when daylight heating of the ground creates rising turbulence which breaks up the surface inversion layer formed in the preceding night by radiation cooling of the ground. This type of fumigation has been categorized as "Type I" and is referred to as "nocturnal inversion breakup fumigation" or simply as "inversion breakup fumigation".

Another type of fumigation occurs at night when stable rural air drifts over an urban area and encounters mechanical turbulence as well as rising thermal turbulence from the urban area's heat emission. A plume embedded in the stable rural air can be fumigated by such an encounter with the urban area turbulence. This type of fumigation has been categorized as "Type II" and is referred to as "nocturnal urban fumigation".

Yet another type of fumigation occurs in shoreline areas and is conceptually similar to the Type II fumigation. When cool, stable air above a lake or ocean blows onshore during the day-time, it encounters rising turbulence from the warm land. The vertical profile boundary between the stable air flowing onshore and the land-based turbulence exhibits a parabolic shape. The boundary starts near ground level at the shoreline and rises upward perhaps 1 to 5 km at downwind onshore distances of 10 to 40 km from the shoreline. Plumes embedded in the stable air (from source stacks located near the shoreline) are fumigated as they encounter the boundary of rising turbulence. Plumes at different heights within the stable air encounter the turbulence boundary at different distances from the shoreline. This type of fumigation has been categorized as "Type III" and is referred to as "continuous or dynamic fumigation".

Lyons has published[23] a remarkably comprehensive treatise which provides an extraordinary insight of shoreline meteorology and of Type III fumigations. Anyone wishing to gain a thorough appreciation of the complexities of predicting stack gas plume dispersion in a shoreline environment would do well to study Lyons' treatise. In fact, anyone wishing to understand the importance of gathering site-specific, real-world meteorological data for stack gas plume dispersion predictions in <u>any environment</u> should study Lyons' treatise.

CLOSING COMMENT ON PREDICTING STACK GAS DISPERSION

This chapter concludes the presentation of the fundamentals of stack gas dispersion. The remainder of the book deals with some associated matters. As the reader has no doubt observed, most of this book has focused upon dispersion predictions for plumes derived from continuous point-source stacks. **It should be emphasized that the current state-of-the-art in air dispersion models goes far beyond the modelling of single point-sources.** There are a variety of much more complex and sophisticated models that are now in use, which have been designed for:

- Predicting short-term (hourly, 3-hourly, and daily) pollutant concentrations from multiple point-sources and multiple area-sources.

- Predicting the long-term (seasonal and annual) pollutant concentrations from multiple point-sources and/or area-sources involving the use of complex source and receptor location grids as well as long-term joint frequency distribution of meteorological data.

- Inclusion of factors such as source stack exit downwash and building wake effects.

- Handling various specific types of terrain such as valley or shoreline environments.

- Inclusion of methods for predicting dispersion during inversion conditions (trapped plumes) and inversion breakups (fumigation).

- Inclusion of chemical reaction, deposition, decay and other removal mechanisms as well as the simple dispersion of plume emissions.

- Predicting the dispersion of motor vehicle emissions.[16]

- Predicting the dispersion of dense gas (heavier-than-air) plumes from non-continuous, emergency releases of hazardous chemicals.[41]

- Many other purposes.

Although the state-of-the-art has progressed well beyond single point-source dispersion models, the reader of this book should now have acquired an excellent foundation in the fundamentals of stack gas dispersion. That foundation will provide the reader with a basis for understanding and evaluating the more complex and more advanced dispersion models. Unless the fundamentals are thoroughly understood, one cannot truly assess the more advanced models or be fully aware of the multitude of assumptions, constraints and limitations involved in their application.

Chapter 10

METEOROLOGICAL DATA

WIND ROSES

Site-specific wind data are often presented graphically in the form of a "wind rose" as typified in Figure 40. A wind rose is a plot of the frequency with which selected categories of windspeed and wind direction occur, and they are based upon historical meteorological records accumulated at a given site.

For example, during the years 1970 to 1974 for the site in Figure 40, southwesterly winds occurred 17.5 percent of the time with this breakdown into various windspeeds:

Southwesterly Windspeeds	Frequency Of Occurrence
1 to 3 miles/hr	5.0 % of the time
4 to 7 miles/hr	6.0 % of the time
8 to 12 miles/hr	6.5 % of the time
1 to 12 miles/hr	17.5 % of the time

As noted by the direction arrows in Figure 40, **wind data and wind roses are almost invariably expressed in terms of the direction from which the wind is coming.** In other words, a southwesterly wind is one which comes from the southwest and blows towards the northeast.

Wind roses can be presented in many different forms, using many different increments or categories of windspeed and wind direction:

WINDSPEEDS may be categorized into given categories by knots, statute miles per hour, or meters per second. Some of the typically used windspeed increments are:

Knots	Statute mi/hr	meters/sec
1-3	1-3	0.5-1.5
4-6	4-7	2.0-3.0
7-10	8-12	3.5-5.0
11-16	13-18	5.5-8.0
17-21	19-24	8.5-11.0
> 21	> 24	> 11

The above categories are approximately equivalent. The exact windspeed equivalents are:

```
        1 knot = 1.152 statute mi/hr = 0.515 m/sec
1 statute mi/hr = 0.868 knots        = 0.447 m/sec
        1 m/sec = 2.237 statute mi/hr = 1.942 knots
```

WIND DIRECTIONS may be categorized into 8, 12, 16 or 36 directions of 45, 30, 22.5 or 10 degrees of angular width, respectively:

```
 8 directions (N, NE, E, SE, S, SW, W, NW)
12 directions (0°, 30°, 60°, 90°, 120°, ... etc.)
16 directions (N, NNE, NE, ENE, E, ... etc. as in Figure 40)
36 directions (0°, 10°, 20°, 30°, 40°, ... etc.)
```

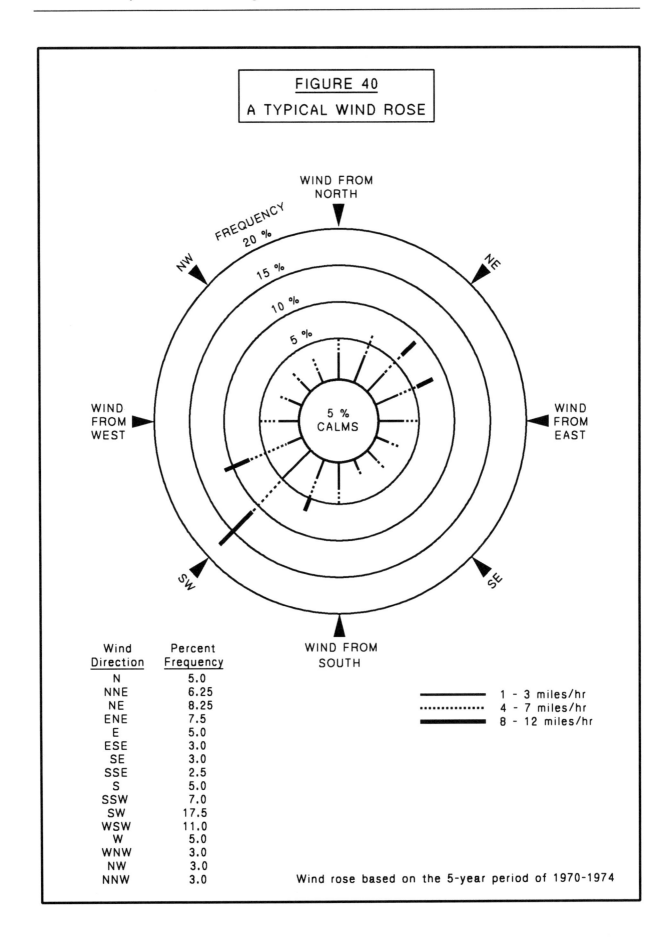

DISTRIBUTION OF CALMS

In most wind data records, time periods with very light winds, or essentially no wind, are designated as "calms". The frequency or occurrence of calms may be presented separately as in Figure 40 (where the calms are noted as occurring 5 percent of the time), or the calms may be distributed within the lowest windspeed category. When the calms are distributed within the lowest windspeed category of 1-3 mi/hr, that category becomes a 0-3 mi/hr category. The wind direction distribution of calms assigned to the lowest windspeed category is usually based upon the wind direction distribution of the <u>two</u> lowest windspeed categories ... so as to use a larger distribution sample of wind direction within the light winds. Mathematically, the distribution of calms can be expressed:

(67) $\quad c_s = (n_s/N)(C) = n_s(C/N)$

> where: N = total frequency of 1-7 mi/hr winds in all direction categories within the two lowest windspeed categories
> n_s = frequency of 1-7 mi/hr winds in any specific direction within the two lowest windspeed categories
> C = total frequency of calms
> c_s = frequency of calms assigned to 0-3 mi/hr winds in any specific direction

Example 9 uses the data from Figure 40 to illustrate how calms are distributed among the wind direction frequencies. The total frequency of calms in Figure 40 is 5 percent. That 5 percent is broken down into the specific wind direction frequencies within what is to be a new, calculated 0-3 mi/hr windspeed category using equation (67). Those specific wind direction frequencies (c_s values) are added to the 1-3 mi/hr wind direction frequencies to obtain the wind direction frequency distribution of the new, calculated 0-3 mi/hr windspeed category.

THE STAR DATA

The collection center and custodian of all United States weather records is the National Climatic Center (NCC) located in Asheville, North Carolina. In addition to its routine climatological services, the NCC in cooperation with the EPA has developed computer programs to provide the meteorological data needed for input to stack gas dispersion models. One of those programs provides site-specific arrays of windspeed and direction frequencies as a function of the Pasquill stability classes, and is referred to as the STAR program (STability ARray).

The STAR data is made available in a number of user-selected optional formats as shown in Table 19:

- The regular STAR data with options for 5, 6 or 7 stability classes.
- The day/night STAR data with options for 6, 7 or 8 stability classes.
- Options for either seasonal or annual data, with either the regular or the day/night STAR data.

The only difference between the regular and the day/night options is that Pasquill's neutral

Chapter 10: Meteorological Data

EXAMPLE 9: DISTRIBUTION OF CALMS USING DATA FROM FIGURE 40

Read from Figure 40, and tabulate below, all the specific wind direction frequency data for the 1-3 mi/hr windspeed category.

Read from Figure 40, and tabulate below, all the specific wind direction frequency data for the 4-7 mi/hr windspeed category.

Add the specific wind direction frequencies of the 1-3 mi/hr winds to those of the 4-7 mi/hr winds to obtain and tabulate the 1-7 mi/hr specific wind direction frequencies. Their sum (of 78.75 percent) is N as defined in equation (67).

The total frequency of calms of 5 percent in Figure 40 is C as defined in equation (67). Thus:

$$C/N = 5/78.75 = 0.0635$$

Multiply the 1-7 mi/hr specific wind direction frequencies, defined as n_s in equation (67), by 0.0635 to obtain the wind direction frequencies of the calms assigned to the 0-3 mi/hr windspeed category. These are the c_s values defined by equation (67).

Add the c_s values to the 1-3 mi/hr specific wind direction frequencies to obtain the calculated wind direction frequencies for the new 0-3 mi/hr category.

DIRECTION	1-3 (mph)		4-7 (mph)		1-7 (mph)		C/N		c_s				0-3 (mph)
N	3.00	+	2.00	=	5.00	×	0.6035	=	0.32	+	3.00	=	3.32
NNE	3.50		2.75		6.25		0.0635		0.40		3.50		3.90
NE	3.00		3.00		6.00		0.0635		0.38		3.00		3.38
ENE	3.00		2.50		5.50		0.0635		0.35		3.00		3.35
E	3.00		2.00		5.00		0.0635		0.32		3.00		3.32
ESE	1.75		1.25		3.00		0.0635		0.19		1.75		1.94
SE	1.75		1.25		3.00		0.0635		0.19		1.75		1.94
SSE	1.50		1.00		2.50		0.0635		0.16		1.50		1.66
S	3.00		2.00		5.00		0.0635		0.32		3.00		3.32
SSW	2.50		2.50		5.00		0.0635		0.32		2.50		2.82
SW	5.00		6.00		11.00		0.0635		0.70		5.00		5.70
WSW	2.00		5.50		7.50		0.0635		0.48		2.00		2.48
W	2.50		2.50		5.00		0.0635		0.32		2.50		2.82
WNW	1.75		1.25		3.00		0.0635		0.19		1.75		1.94
NW	1.75		1.25		3.00		0.0635		0.19		1.75		1.94
NNW	1.50		1.50		3.00		0.0635		0.19		1.50		1.69
	40.50		38.25		78.75				5.00				45.50

The far right column in the above tabulation contains the specific wind direction frequencies of the new, calculated 0-3 mi/hr windspeed category, including the distributed calms.

TABLE 19

OPTIONAL FORMATS OF THE STAR DATA

(available from the National Climatic Center)

	Regular Star Options			Day/Night Star Options		
	7 Stability Classes	6 Stability Classes	5 Stability Classes	8 Stability Classes	7 Stability Classes	6 Stability Classes
EXTREMELY UNSTABLE	A	A	A	A	A	A
UNSTABLE	B	B	B	B	B	B
SLIGHTLY STABLE	C	C	C	C	C	C
NEUTRAL/DAY				D	D	D
NEUTRAL	D	D	D			
NEUTRAL NIGHT				E	E	E
SLIGHTLY STABLE	E	E	E	F	F	
STABLE	F	F		G	G	F
EXTREMELY STABLE	G			H		

For any of the above options, the STAR outputs are available in either one of these sub-options:

- An annual basis
- A seasonal basis such as:
  - DJF --- December, January, February (Winter)
  - MAM --- March, April, May (Spring)
  - JJA --- June, July, August (Summer)
  - SON --- September, October, November (Fall)

Assuming that the STAR data is derived from 8 observations per day over a total period of 5-years:

- The annual basis output would be derived from:
  $(5)(365)(8) = 14,600$ observations

- The seasonal basis outputs would be derived from:
  - DJF = $(90/365)(14,600)$ = 3,600 observations
  - MAM = $(92/365)(14,600)$ = 3,680 observations
  - JJA = $(92/365)(14,600)$ = 3,680 observations
  - SON = $(91/365)(14,600)$ = 3,640 observations
  -                                    14,600

stability class used in the regular STAR data is split into day-time and night-time neutral classes for the day/night option. Hence, each day/night option categorizes the meteorological data into one more stability class than does the counterpart regular option.

As discussed in Chapter 1, the Pasquill stability classes can be defined by a combination of the windspeed, the amount of cloud cover and a subjective estimate of the solar radiation insolation. In the STAR data, the stability classes have been determined by an objective method of defining those characteristics, developed by Turner,[62] which lends itself to computer calculation of the stability classes. Turner's objective method, as used in the STAR program, is outlined in Table 20 and summarized below:

- For day-time observations, the insolation class (IC) is defined as a function of the solar altitude at the time and location of the observation.

- For day-time observations, a net radiation index (NRI) is then determined as a function of:

    -- the insolation class (IC),
    -- the percentage cloud cover, and
    -- the cloud ceiling height.

    For night-time observations, the net radiation index (NRI) is determined as a function of:

    -- the percentage cloud cover and
    -- the cloud ceiling height.

- The Pasquill stability class is then determined as a function of the net radiation index (NRI) and the windspeed category at the time and location of the observation.

The data basis and the determination of the stability classes is an integral part of the STAR program. The user of the STAR program output sees only the resulting data arrays in one of the user-selected optional formats explained in Table 19.

A typical set of STAR output data is presented in Table 21. The selected options and other specifics of the data set in Table 21 are:

Type	Day/Night STAR
Stability classes	6
Period	Annual
Base period	5-years, 1973-1977
Observations/day	8
Total observations	14,600
Station	South Bend, Indiana

The data for each stability class is arrayed first as the frequency distribution of each combination of windspeed and wind direction in that stability class. The data is then arrayed as the relative frequency distribution of each combination of windspeed and wind direction in that same stability class. Thus, the data for each stability class consists of two arrays, one for the frequency distributions and one for the relative frequency distributions. To clarify

TABLE 20

OBJECTIVE DEFINITION OF STABILITY CLASSES

DAY-TIME INSOLATION CLASS		
SOLAR ALTITUDE (A)	INSOLATION	INSOLATION CLASS (IC)
A > 60°	Strong	4
35° < A ≤ 60°	Moderate	3
15° < A ≤ 35°	Slight	2
A ≤ 15°	Weak	1

NET RADIATION INDEX			
PERCENT CLOUD COVER (B)	CEILING (C), ft	TIME	NET RADIATION INDEX (NRI)
B ≤ 50	(any ceiling)	day	NRI = IC
100 > B > 50	C < 7000	day	NRI = IC - 2 [†]
B = 100	C < 7000	day	NRI = 0
B = 100	C ≥ 7000	day	NRI = IC - 1 [†]
100 > B > 50	16000 > C ≥ 7000	day	NRI = IC - 1 [†]
100 > B > 50	C ≥ 16000	day	NRI = IC
B ≤ 40	(any ceiling)	night	NRI = -2
B = 100	C < 7000	night	NRI = 0
100 > B > 40	C < 7000	night	NRI = -1
B > 40	C ≥ 7000	night	NRI = -1

[†] If (IC - 2) or (IC - 1) is less than 1, let NRI = 1

PASQUILL STABILITY CLASS AS A FUNCTION OF NET RADIATION INDEX AND WINDSPEED							
Windspeed (knots)	NET RADIATION INDEX						
	4	3	2	1	0	-1	-2
0, 1	EU	EU	U	SU	N	S	ES
2, 3	EU	U	U	SU	N	S	ES
4, 5	EU	U	SU	N	N	SS	S
6	U	U	SU	N	N	SS	S
7	U	U	SU	N	N	N	SS
8, 9	U	SU	SU	N	N	N	SS
10	SU	SU	N	N	N	N	SS
11	SU	SU	N	N	N	N	N
≥ 12	SU	N	N	N	N	N	N

EU = extremely unstable   U = unstable
SU = slightly unstable   N = neutral
SS = slightly stable   S = stable
ES = extremely stable

(Multiply knots by 1.152 to obtain statute miles/hr)

the distribution terminology:

Frequency Distribution: These arrays present the **number of observed occurrences** of specific windspeed and direction combinations during specific stability occurrences over the entire 5-year period. For example, in Table 21, the total observed occurrences of C stability during the 5-year period was 1,353 ... and the total observed occurrences of 7-10 knot winds from the SW during C stability conditions was 72 during the same 5-year period.

Relative Frequency Distribution: These arrays present the occurrences of specific windspeed and direction combinations during specific stability occurrences as a **fraction of the total observations** over the entire 5-year period. Since the total observations were 14,600 over the 5-year period, the fractional occurrence (i.e. relative frequency) of C stabilities during that period was 1,353/14,600 = 0.092761 ... and the relative frequency of 7-10 knot winds from the SW during C stability conditions was only 72/14,600 = 0.004932. (As a note of interest, the relative frequencies can be multiplied by 100 to obtain the percentage frequency for any particular combination of windspeed, wind direction and stability class).

The frequency distribution (i.e., observed occurrences) array for each stability class lists the calms separately and the lowest windspeed category is noted as 1-3 knots. However, the relative frequency distribution array for each stability class includes the calms distributed into the lowest windspeed category which is then noted as being 0-3 knots.

These are some examples of how to use the STAR data as it is typified in Table 21:

QUESTION: How often did 7-10 knot SE winds occur during the total 5-year period?
ANSWER: The last page in Table 21 presents windspeed and direction data on an aggregated basis for all stability conditions. From the frequency distributions in the upper part of that page, such winds occurred 270 times during the 5-year period. Alternatively, from the relative frequency distributions in the lower part of the page, such winds occurred 0.018493(14,600) = 270 times.

QUESTION: How often can 7-10 knot SE winds be expected to occur in any given year?
ANSWER: 270/5 = 54 times per year.

QUESTION: How often can 7-10 knot SE winds be expected to occur during D stability conditions in any given year?
ANSWER: From the frequency distributions (upper page part) of the D stability page in Table 21, such winds can be expected to occur 74/5 = 14.8 times per year. Alternatively, from the relative frequency distributions (lower page part), such winds can be expected 0.005068(14,600)/5 = 14.8 times per year.

QUESTION: How often can D-stability conditions be expected during any given year?
ANSWER: From the upper part of the D stability page in Table 21, D stability conditions can be expected 4,090/5 = 818 times per year. Alternatively, from the lower part of the D stability page, D-stability conditions can be expected 0.280137(14,600)/5 = 818 times per year.

Additional information on the STAR program is available from the National Climatic Center.[63]

TABLE 21

A TYPICAL SET OF STAR DATA

(contained in the next six pages)

Type	Day/Night Star
Stability classes	6
Period	Annual
Windspeed units	Knots
Wind directions	16
Base period	5 years, 1973-1977
Observations/day	8
Total observations	14,600
Station	South Bend, Indiana

STABILITY A
FREQUENCY DISTRIBUTION

WIND FROM	__1-3__	__4-6__	__7-10__	__11-16__	__17-21__	__>21__	AVG SPEED	TOTAL
N	0	6	0	0	0	0	4.2	6
NNE	1	0	0	0	0	0	3.0	1
NE	0	0	0	0	0	0	0.0	0
ENE	1	1	0	0	0	0	3.5	2
E	2	2	0	0	0	0	3.8	4
ESE	0	2	0	0	0	0	4.0	2
SE	0	2	0	0	0	0	5.0	2
SSE	0	1	0	0	0	0	4.0	1
S	0	2	0	0	0	0	4.0	2
SSW	0	0	0	0	0	0	0.0	0
SW	2	6	0	0	0	0	4.3	8
WSW	1	0	0	0	0	0	3.0	1
W	0	2	0	0	0	0	4.5	2
WNW	0	2	0	0	0	0	4.5	2
NW	1	2	0	0	0	0	4.0	3
NNW	0	0	0	0	0	0	0.0	0
Average	2.9	4.4	0.0	0.0	0.0	0.0	2.9	
Total	8	28	0	0	0	0		

Number of A stability occurrences = 51
Number of A stability calms = 15

RELATIVE FREQUENCY DISTRIBUTION

	__0-3__	__4-6__	__7-10__	__11-16__	__17-21__	__>21__	TOTAL
N	0.000171	0.000411	0.000000	0.000000	0.000000	0.000000	0.000582
NNE	0.000097	0.000000	0.000000	0.000000	0.000000	0.000000	0.000097
NE	0.000000	0.000000	0.000000	0.000000	0.000000	0.000000	0.000000
ENE	0.000126	0.000068	0.000000	0.000000	0.000000	0.000000	0.000194
E	0.000251	0.000137	0.000000	0.000000	0.000000	0.000000	0.000388
ESE	0.000057	0.000137	0.000000	0.000000	0.000000	0.000000	0.000194
SE	0.000057	0.000137	0.000000	0.000000	0.000000	0.000000	0.000194
SSE	0.000029	0.000068	0.000000	0.000000	0.000000	0.000000	0.000097
S	0.000057	0.000137	0.000000	0.000000	0.000000	0.000000	0.000194
SSW	0.000000	0.000000	0.000000	0.000000	0.000000	0.000000	0.000000
SW	0.000365	0.000411	0.000000	0.000000	0.000000	0.000000	0.000776
WSW	0.000097	0.000000	0.000000	0.000000	0.000000	0.000000	0.000097
W	0.000057	0.000137	0.000000	0.000000	0.000000	0.000000	0.000194
WNW	0.000057	0.000137	0.000000	0.000000	0.000000	0.000000	0.000194
NW	0.000154	0.000137	0.000000	0.000000	0.000000	0.000000	0.000291
NNW	0.000000	0.000000	0.000000	0.000000	0.000000	0.000000	0.000000
Total	0.001575	0.001918	0.000000	0.000000	0.000000	0.000000	

Relative frequency of A stability occurrences = 0.003493
Relative frequency of A stability calms distributed above = 0.001027

STABILITY B
FREQUENCY DISTRIBUTION

WIND FROM	WINDSPEED (KNOTS)						AVG SPEED	TOTAL
	1-3	4-6	7-10	11-16	17-21	>21		
N	7	14	12	0	0	0	5.8	33
NNE	2	7	10	0	0	0	6.4	19
NE	4	12	7	0	0	0	5.7	23
ENE	7	11	7	0	0	0	5.2	25
E	13	14	9	0	0	0	5.1	36
ESE	2	8	9	0	0	0	6.2	19
SE	6	15	7	0	0	0	5.4	28
SSE	2	11	5	0	0	0	5.4	18
S	8	19	15	0	0	0	5.7	42
SSW	7	13	8	0	0	0	5.4	28
SW	7	12	9	0	0	0	5.5	28
WSW	3	19	21	0	0	0	6.2	43
W	5	17	14	0	0	0	5.8	36
WNW	2	12	12	0	0	0	6.3	26
NW	2	20	12	0	0	0	6.0	34
NNW	2	13	4	0	0	0	5.5	19
Average	2.9	5.3	7.6	0.0	0.0	0.0	4.8	
Total	79	217	161	0	0	0		

Number of B stability occurrences = 546
Number of B stability calms = 89

RELATIVE FREQUENCY DISTRIBUTION

	0-3	4-6	7-10	11-16	17-21	>21	TOTAL
N	0.000912	0.000959	0.000822	0.000000	0.000000	0.000000	0.002693
NNE	0.000322	0.000479	0.000685	0.000000	0.000000	0.000000	0.001487
NE	0.000603	0.000822	0.000479	0.000000	0.000000	0.000000	0.001905
ENE	0.000850	0.000753	0.000479	0.000000	0.000000	0.000000	0.002083
E	0.001446	0.000959	0.000616	0.000000	0.000000	0.000000	0.003022
ESE	0.000343	0.000548	0.000616	0.000000	0.000000	0.000000	0.001507
SE	0.000843	0.001207	0.000479	0.000000	0.000000	0.000000	0.002350
SSE	0.000405	0.000753	0.000342	0.000000	0.000000	0.000000	0.001501
S	0.001104	0.001301	0.001027	0.000000	0.000000	0.000000	0.003433
SSW	0.000891	0.000890	0.000548	0.000000	0.000000	0.000000	0.002330
SW	0.000871	0.000822	0.000616	0.000000	0.000000	0.000000	0.002309
WSW	0.000659	0.001301	0.001438	0.000000	0.000000	0.000000	0.003398
W	0.000796	0.001164	0.000959	0.000000	0.000000	0.000000	0.002919
WNW	0.000425	0.000822	0.000822	0.000000	0.000000	0.000000	0.002069
NW	0.000590	0.001370	0.000822	0.000000	0.000000	0.000000	0.002782
NNW	0.000446	0.000890	0.000274	0.000000	0.000000	0.000000	0.001610
Total	0.011507	0.014863	0.011027	0.000000	0.000000	0.000000	

Relative frequency of B stability occurrences = 0.037397
Relative frequency of B stability calms distributed above = 0.006096

STABILITY C
FREQUENCY DISTRIBUTION

WIND FROM	_____ WINDSPEED (KNOTS) _____						AVG SPEED	TOTAL
	1-3	4-6	7-10	11-16	17-21	>21		
N	2	39	62	15	3	0	8.0	121
NNE	1	15	23	5	0	0	7.5	44
NE	0	18	22	6	0	0	7.6	46
ENE	2	11	29	3	1	0	7.7	46
E	0	31	45	3	0	0	7.3	79
ESE	0	21	23	2	0	0	7.0	46
SE	2	14	31	7	0	0	7.7	54
SSE	0	14	40	7	0	0	8.0	61
S	3	38	74	19	0	0	7.9	134
SSW	3	31	75	10	0	0	7.9	119
SW	3	38	72	24	4	0	8.4	141
WSW	3	46	71	27	4	0	8.6	151
W	2	19	34	24	5	0	9.2	84
WNW	0	16	36	17	3	0	9.2	72
NW	1	13	36	19	0	0	9.0	69
NNW	2	14	25	14	1	0	8.7	56
Average	2.9	5.1	8.6	12.4	17.9	0.0	8.0	
Total	24	378	698	202	21	0		

Number of C stability occurrences = 1353
Number of C stability calms = 30

RELATIVE FREQUENCY DISTRIBUTION

	0-3	4-6	7-10	11-16	17-21	>21	TOTAL
N	0.000347	0.002671	0.004247	0.001027	0.000205	0.000000	0.008497
NNE	0.000150	0.001027	0.001575	0.000342	0.000000	0.000000	0.003095
NE	0.000092	0.001233	0.001507	0.000411	0.000000	0.000000	0.003243
ENE	0.000203	0.000753	0.001986	0.000205	0.000068	0.000000	0.003217
E	0.000158	0.002123	0.003082	0.000205	0.000000	0.000000	0.005569
ESE	0.000107	0.001438	0.001575	0.000137	0.000000	0.000000	0.003258
SE	0.000219	0.000959	0.002123	0.000479	0.000000	0.000000	0.003780
SSE	0.000072	0.000959	0.002740	0.000479	0.000000	0.000000	0.004250
S	0.000415	0.002603	0.005068	0.001301	0.000000	0.000000	0.009388
SSW	0.000379	0.002123	0.005137	0.000685	0.000000	0.000000	0.008324
SW	0.000415	0.002603	0.004932	0.001644	0.000274	0.000000	0.009867
WSW	0.000456	0.003151	0.004863	0.001849	0.000274	0.000000	0.010593
W	0.000244	0.001301	0.002329	0.001644	0.000342	0.000000	0.005861
WNW	0.000082	0.001096	0.002466	0.001164	0.000205	0.000000	0.005013
NW	0.000140	0.000890	0.002466	0.001301	0.000000	0.000000	0.004798
NNW	0.000219	0.000959	0.001712	0.000958	0.000068	0.000000	0.003917
Total	0.003699	0.025890	0.047808	0.013836	0.001438	0.000000	

Relative frequency of C stability occurrences = 0.092671
Relative frequency of C stability calms distributed above = 0.002055

STABILITY D
FREQUENCY DISTRIBUTION

WIND FROM	WINDSPEED (KNOTS)						AVG SPEED	TOTAL
	1-3	4-6	7-10	11-16	17-21	>21		
N	1	43	73	126	18	4	11.1	265
NNE	0	24	36	28	4	0	9.5	92
NE	2	33	35	33	4	0	9.1	106
ENE	0	21	31	32	7	0	10.1	91
E	1	53	78	82	6	0	9.6	220
ESE	2	31	56	78	12	2	10.9	181
SE	2	27	74	91	9	4	10.8	207
SSE	0	21	53	64	5	1	10.5	144
S	1	59	134	169	19	4	10.8	386
SSW	4	33	86	147	33	11	11.9	314
SW	1	44	128	237	80	26	13.0	516
WSW	3	59	91	189	70	12	12.5	424
W	2	32	63	130	44	22	13.0	293
WNW	4	17	62	122	42	5	12.6	252
NW	3	28	74	214	40	2	12.0	361
NNW	2	14	58	112	19	2	12.0	207
Average	3.0	5.1	8.9	13.3	18.3	24.4	11.6	
Total	28	539	1132	1854	411	95		

Number of D stability (neutral/day) occurrences = 4090
Number of D stability (neutral/day) calms = 31

RELATIVE FREQUENCY DISTRIBUTION

	0-3	4-6	7-10	11-16	17-21	>21	TOTAL
N	0.000233	0.002945	0.005000	0.008630	0.001233	0.000274	0.018315
NNE	0.000090	0.001644	0.002466	0.001918	0.000274	0.000000	0.006391
NE	0.000628	0.002260	0.002397	0.002260	0.000205	0.000000	0.007391
ENE	0.000079	0.001438	0.002123	0.002192	0.000479	0.000000	0.006312
E	0.000271	0.003630	0.005342	0.005616	0.000411	0.000000	0.015271
ESE	0.000261	0.002123	0.003836	0.005342	0.000822	0.000137	0.012521
SE	0.000246	0.001849	0.005068	0.006233	0.000616	0.000274	0.014287
SSE	0.000079	0.001438	0.003630	0.004384	0.000342	0.000068	0.009942
S	0.000293	0.004041	0.009178	0.011575	0.001301	0.000274	0.026663
SSW	0.000413	0.002260	0.005890	0.010068	0.002260	0.000753	0.021645
SW	0.000237	0.003014	0.008767	0.016233	0.005479	0.001781	0.035511
WSW	0.000438	0.004041	0.006233	0.012945	0.004795	0.000822	0.029273
W	0.000264	0.002192	0.004315	0.008904	0.003014	0.001507	0.020196
WNW	0.000353	0.001164	0.004247	0.008356	0.002877	0.000342	0.017339
NW	0.000322	0.001918	0.005068	0.014658	0.002740	0.000137	0.024842
NNW	0.000197	0.000959	0.003973	0.007671	0.001301	0.000137	0.014238
Total	0.004041	0.036918	0.077534	0.126986	0.028151	0.006507	

Relative frequency of D stability (neutral/day) occurrences = 0.280137
Relative frequency of D stability (neutral/day) calms distributed above = 0.002123

STABILITY E
FREQUENCY DISTRIBUTION

WIND FROM	_____ WINDSPEED (KNOTS) _____						AVG SPEED	TOTAL
	1-3	4-6	7-10	11-16	17-21	>21		
N	11	60	129	93	15	0	9.4	308
NNE	4	33	46	29	3	0	8.5	115
NE	5	38	62	47	2	0	8.9	154
ENE	1	37	53	33	3	0	9.0	127
E	4	39	125	75	14	1	9.8	258
ESE	1	42	119	125	13	0	10.4	300
SE	4	16	123	105	13	0	10.7	261
SSE	0	21	92	60	6	1	9.9	180
S	7	53	225	168	14	2	10.0	469
SSW	1	30	189	170	19	3	10.7	412
SW	6	65	238	300	51	7	11.2	667
WSW	8	53	168	251	51	20	11.8	551
W	5	43	98	154	48	24	12.4	372
WNW	4	32	90	139	29	6	11.5	300
NW	6	21	99	105	23	3	11.1	257
NNW	5	26	61	59	12	0	10.3	163
Average	3.0	5.1	8.6	13.1	18.2	24.3	10.5	
Total	72	609	1917	1913	316	67		

Number of E stability (neutral/night) occurrences = 4997
Number of E stability (neutral/night) calms = 103

RELATIVE FREQUENCY DISTRIBUTION

	0-3	4-6	7-10	11-16	17-21	>21	TOTAL
N	0.001489	0.004110	0.008836	0.006370	0.001027	0.000000	0.021831
NNE	0.000657	0.002260	0.003151	0.001986	0.000205	0.000000	0.008260
NE	0.000788	0.002603	0.004247	0.003219	0.000137	0.000000	0.010993
ENE	0.000462	0.002534	0.003630	0.002260	0.000205	0.000000	0.009092
E	0.000719	0.002671	0.008562	0.005137	0.000959	0.000068	0.018117
ESE	0.000514	0.002877	0.008151	0.008562	0.000890	0.000000	0.020993
SE	0.000481	0.001096	0.008425	0.007192	0.000890	0.000000	0.018084
SSE	0.000218	0.001438	0.006301	0.004110	0.000411	0.000068	0.012546
S	0.001101	0.003630	0.015411	0.011507	0.000959	0.000137	0.032745
SSW	0.000390	0.002055	0.012945	0.011644	0.001301	0.000205	0.028540
SW	0.001146	0.004452	0.016301	0.020548	0.003493	0.000479	0.046420
WSW	0.001180	0.003630	0.011507	0.017192	0.003493	0.001370	0.038372
W	0.000840	0.029454	0.006712	0.010548	0.003288	0.001644	0.025977
WNW	0.000647	0.002192	0.006164	0.009521	0.001986	0.000411	0.020921
NW	0.000691	0.001438	0.006781	0.007192	0.001575	0.000205	0.017882
NNW	0.000664	0.001781	0.004178	0.004041	0.000822	0.000000	0.011486
Total	0.011986	0.041712	0.131301	0.131027	0.021644	0.004589	

Relative frequency of E stability (neutral/night) occurrences = 0.342260
Relative frequency of E stability (neutral/night) calms distributed above = 0.007055

STABILITY F
FREQUENCY DISTRIBUTION

WIND FROM	WINDSPEED (KNOTS)						AVG SPEED	TOTAL
	1-3	4-6	7-10	11-16	17-21	>21		
N	47	192	89	0	0	0	5.6	328
NNE	18	88	10	0	0	0	4.8	116
NE	8	84	15	0	0	0	5.2	107
ENE	17	94	15	0	0	0	5.0	126
E	23	117	25	0	0	0	5.1	165
ESE	7	84	36	0	0	0	5.9	127
SE	8	56	35	0	0	0	6.2	99
SSE	7	59	26	0	0	0	5.9	92
S	27	205	94	0	0	0	5.9	326
SSW	14	129	86	0	0	0	6.3	229
SW	22	193	202	0	0	0	6.7	417
WSW	30	190	70	0	0	0	5.6	290
W	36	117	50	0	0	0	5.4	203
WNW	13	77	32	0	0	0	5.6	122
NW	18	84	35	0	0	0	5.6	137
NNW	19	90	38	0	0	0	5.6	147
Average	3.0	5.0	8.4	0.0	0.0	0.0	4.9	
Total	314	1859	858	0	0	0		

Number of F stability occurrences = 3653
Number of F stability calms = 532

RELATIVE FREQUENCY DISTRIBUTION

	0-3	4-6	7-10	11-16	17-21	>21	TOTAL
N	0.007227	0.013151	0.006096	0.000000	0.000000	0.000000	0.026473
NNE	0.003010	0.006027	0.000685	0.000000	0.000000	0.000000	0.009723
NE	0.002091	0.005753	0.002027	0.000000	0.000000	0.000000	0.008871
ENE	0.003026	0.006438	0.001027	0.000000	0.000000	0.000000	0.010491
E	0.003923	0.008014	0.001712	0.000000	0.000000	0.000068	0.013649
ESE	0.002005	0.005753	0.002466	0.000000	0.000000	0.000000	0.010225
SE	0.001621	0.003836	0.002397	0.000000	0.000000	0.000000	0.007854
SSE	0.001586	0.004041	0.001781	0.000000	0.000000	0.000000	0.007408
S	0.005740	0.014041	0.006438	0.000000	0.000000	0.000000	0.026219
SSW	0.003357	0.008836	0.005890	0.000000	0.000000	0.000000	0.018083
SW	0.005112	0.013219	0.013836	0.000000	0.000000	0.000000	0.032167
WSW	0.005744	0.013014	0.004795	0.000000	0.000000	0.000000	0.023552
W	0.005031	0.008014	0.003425	0.000000	0.000000	0.000000	0.016470
WNW	0.002400	0.005274	0.002192	0.000000	0.000000	0.000000	0.009865
NW	0.002943	0.005753	0.002397	0.000000	0.000000	0.000000	0.011094
NNW	0.003129	0.006164	0.002603	0.000000	0.000000	0.000000	0.011896
Total	0.057945	0.127329	0.058767	0.000000	0.000000	0.000000	

Relative frequency of F stability occurrences = 0.244041
Relative frequency of F stability calms distributed above = 0.036438

ALL STABILITIES AGGREGATED
FREQUENCY DISTRIBUTION

WIND FROM	WINDSPEED (KNOTS)						AVG SPEED	TOTAL
	1-3	4-6	7-10	11-16	17-21	>21		
N	68	354	365	234	36	4	8.3	1061
NNE	26	167	125	62	7	0	7.4	387
NE	19	185	141	86	5	0	7.7	436
ENE	28	175	135	68	11	0	7.6	417
E	43	256	282	160	20	1	8.2	762
ESE	12	188	243	205	25	2	9.3	675
SE	22	130	270	203	22	4	9.6	651
SSE	9	127	216	131	11	2	8.9	496
S	46	376	542	356	33	6	8.9	1359
SSW	29	236	444	327	52	14	9.7	1102
SW	41	358	649	561	135	33	10.4	1777
WSW	48	367	421	467	125	32	10.2	1460
W	50	230	259	308	97	46	10.6	990
WNW	23	156	232	278	74	11	10.5	774
NW	31	168	256	338	63	5	10.4	861
NNW	30	157	186	185	32	2	9.4	592
Average	3.0	5.1	8.6	13.1	18.3	24.4	9.0	
Total	525	3630	4766	3969	748	162		

Total number of observations (during the 5-year period) = 14,600
Total number of calms (during the 5-year period) = 800

RELATIVE FREQUENCY DISTRIBUTION

	0-3	4-6	7-10	11-16	17-21	>21	TOTAL
N	0.010223	0.024247	0.025000	0.016027	0.002466	0.000277	0.078236
NNE	0.004326	0.011438	0.008562	0.004247	0.000479	0.000000	0.029052
NE	0.003992	0.012671	0.009658	0.005890	0.000342	0.000000	0.032553
ENE	0.004595	0.011986	0.009247	0.004658	0.000753	0.000000	0.031239
E	0.006888	0.017534	0.019315	0.010959	0.001370	0.000068	0.056135
ESE	0.003459	0.012877	0.016644	0.014041	0.001712	0.000137	0.048870
SE	0.003511	0.008904	0.018493	0.013904	0.001507	0.000274	0.046594
SSE	0.002410	0.008699	0.014795	0.008973	0.000753	0.000137	0.035766
S	0.008716	0.025753	0.037123	0.024384	0.002260	0.000411	0.098467
SSW	0.005481	0.016164	0.030411	0.022397	0.003562	0.000959	0.078974
SW	0.008070	0.024521	0.044452	0.038425	0.009247	0.002260	0.126974
WSW	0.008761	0.025137	0.028836	0.031986	0.008562	0.002192	0.105473
W	0.007117	0.015753	0.017740	0.021096	0.006644	0.003151	0.071501
WNW	0.003936	0.010685	0.015890	0.019041	0.005068	0.000753	0.055374
NW	0.004748	0.011507	0.017534	0.023151	0.004315	0.000342	0.061597
NNW	0.004521	0.010753	0.012740	0.012671	0.002192	0.000137	0.043014
Total	0.090753	0.248630	0.326438	0.271849	0.051233	0.011096	

Relative frequency of observations (during the 5-year period) = 1.000000
Relative frequency of calms (during the 5-year period) distributed above = 0.054795

Chapter 11

FLARE STACK PLUME RISE

INTRODUCTION

Flare stacks are widely used in industrial plants for the disposal of combustible vent gases which are released during:

- Continuous, routine plant operations
- Scheduled plant maintenance periods, when pressure vessels are depressured
- Unscheduled, emergency venting

In particular, flare stacks are an essential safety requirement in hydrocarbon processing facilities such as petroleum refineries, petrochemical plants, natural gas processing and treating plants, hydrocarbon storage spheres and tanks, onshore and offshore oil and gas production facilities, etc.

A flare stack is a vent gas stack with a small pilot flame at the stack exit. Combustible vent gases flowing from the stack exit are ignited by the pilot flame and burned in the open atmosphere just above the stack exit. The hot, combusted gas plume then rises and disperses in the atmosphere just as does any hot, buoyant plume.

A typical flare stack in a large hydrocarbon processing plant may be capable of handling flames with combustion heat releases (Q_c) of as much as 30×10^9 Btu/hr.[†] Such flare stacks may range up to 3-5 feet in diameter, 100-350 feet in height, and have exit gas velocities ranging up to as high as 600 feet/second.

The vent gases sent through a flare stack must have enough fuel content (i.e., sufficient heating value) to support self-combustion at the stack exit, since the only auxiliary fuel supplied to the flare stack is that needed to maintain the small pilot flame. Vent gases which cannot support self-combustion are usually incinerated rather than flared. By definition, this chapter deals only with vent gases which can support self-combustion and are sent through flare stacks.

The purpose of this chapter is to discuss and define the parameters involved in determining the plume rise of the hot, combusted gases from a flare stack flame. Figure 41 depicts a flare stack flame and defines some of the key terminology used in this chapter:

- The flared gas (i.e., the fuel for the flame) is denoted as FG.
- The combusted gas, which results from burning of the flared gas and entrained air, is denoted as CG.
- The visible flame length is denoted as L.
- The stack height is denoted as h_s.

[†] Based on the net heating value of the flared gases. Not to be confused with the sensible heat emission term Q in Briggs' plume rise equations.

Chapter 11: Flare Stack Plume Rise

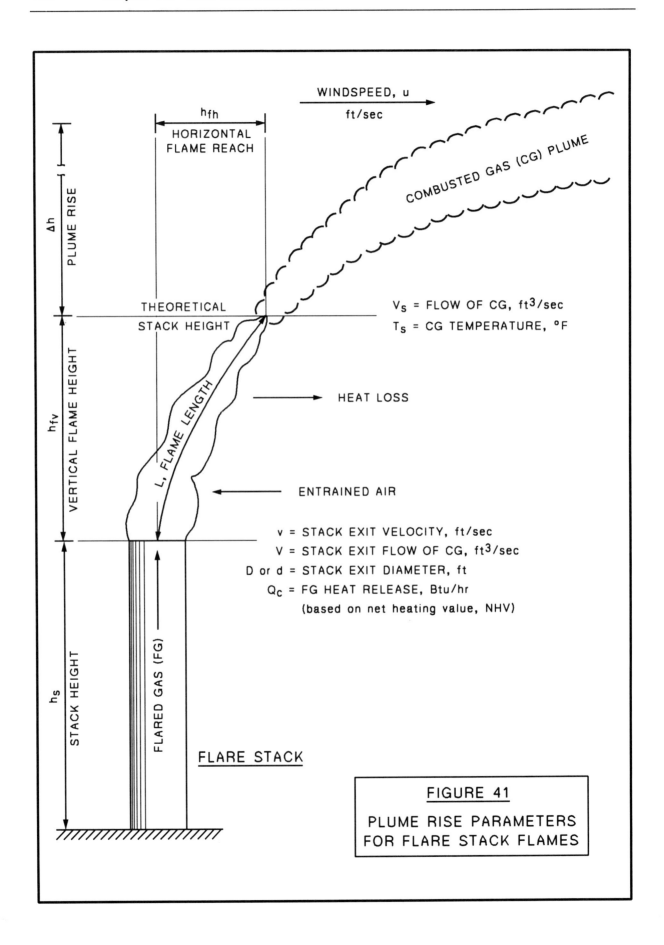

FIGURE 41
PLUME RISE PARAMETERS FOR FLARE STACK FLAMES

- The vertical height vector of the visible flame is denoted as h_{fv}, and the horizontal downwind vector (i.e., the flame "reach") is denoted as h_{fh}.
- The plume rise of the hot combusted gas plume is denoted as Δh and is defined as starting at the end of the visible flame.
- The heat release during the combustion of the flared gas is denoted as Q_c, and is based on the net heating value (NHV) of the flared gas.
- The "theoretical" stack height is defined as the sum of the stack height (h_s) and the vertical height vector of the visible flame (h_{fv}). Thus, the plume centerline height or effective stack height (i.e., H_e) for the combusted gas plume is:

$$H_e = h_s + h_{fv} + \Delta h$$

FLAME LENGTH

As mentioned above and noted in Figure 41, the plume rise (Δh) of the hot, combusted gas plume is defined as starting at the end of the visible flame. That is an arbitrary definition based upon this reasoning:

- A hydrocarbon gas burning under stoichiometric conditions[†] and with no heat lost or transferred from the flame (i.e., adiabatic conditions), will have an adiabatic flame temperature of 3500 °F or more.

- The Briggs plume rise equations (see Chapter 4) were largely derived using plume rise data from flue gas stacks for combustion furnaces such as steam-generating boilers. In such boilers, 85 percent or more of the combustion heat release is transferred into the generation of steam and, thus, the resulting temperature of the combusted gas from the boiler stacks is typically within the range of 250 to 500 °F. The Briggs plume rise equations are best applied to plumes within that temperature range or fairly close to it.

    In the opinion of this author, it would be inappropriate to use Briggs' equations for calculating a flare stack plume rise if the plume rise were defined as starting within the flame center where the temperatures may be 3500 °F or more.

- As the flare stack flame burns, it entrains air from the surrounding atmosphere to supply the oxygen needed for combustion. In fact, the flame entrains excess air, over and above that needed for combustion. The entrainment of excess air increases the volume and lowers the temperature of the combusted gas. Heat lost from the flame by thermal radiation also lowers the combusted gas temperature. The cumulative effect, as discussed later in this chapter, **is a combusted gas temperature at the end of the visible flame that is within the range of 1200 to 1800 °F.** At those temperatures, the Briggs plume rise equations can be used with more confidence than at the flame center temperatures. For that reason, the combusted gas plume rise from a flare stack is defined as starting at the end of the visible flame.

Having defined the end or the tip of the flame as the starting point for the flare stack plume rise equations, some method of estimating the flame length is needed to establish the vertical

---

[†] Where the exact amount of air needed for complete combustion is supplied.

and horizontal location of the flame tip.

An API publication[64] provides a plot of flame length as a function of the flared gas heat release obtained from large-scale, flare stack tests. The API data are re-plotted in Figure 42. The API's correlation line for their data can be expressed as:

(68) $\qquad L = 0.006 \, Q_c^{0.478}$

$\qquad$ where: $\quad L$ = flame length, ft

$\qquad\qquad\qquad Q_c$ = flared gas heat release, Btu/hr

Straitz[65] published similar flame length data obtained from some small-scale, flare stack tests. The Straitz data points are included in Figure 42 and are in good agreement with the API's correlation line.

Steward[66] developed a theoretical mathematical model for turbulent diffusion flames. He then correlated a large body of experimental flame length data against the key parameters indicated by his theoretical model. Unfortunately, Steward's publication did not include the experimental flame length data points. Thus, we can only examine Steward's correlating equation:

(69) $\qquad L/R = 16.218 \, N^{0.2}$

$\qquad$ where: $\quad L$ = flame length, ft

$\qquad\qquad\qquad R$ = stack exit radius, ft

$\qquad\qquad\qquad N$ = a combustion parameter

$\qquad\qquad\qquad\quad = [Q_c^2(r + w\rho_a/\rho)^2] / [R^5(NHV)^2 \rho_a^2 \, g \, (1 - w)^5]$

$\qquad\qquad\qquad r$ = stoichiometric air:fuel ratio, lbs air/lb of flared gas

$\qquad\qquad\qquad w$ = a combustion parameter

$\qquad\qquad\qquad\quad = (rc_pT_a) / (rc_pT_a + NHV)$

$\qquad\quad NHV$ = flared gas net heating value, Btu/lb

$\qquad\qquad\qquad g$ = gravitational constant of 417×10^6 ft/hr^2

$\qquad\qquad\qquad \rho_a$ = ambient air density, lbs/ft^3

$\qquad\qquad\qquad \rho$ = fuel density, lbs/ft^3

$\qquad\qquad\qquad T_a$ = air temperature, °R

$\qquad\qquad\qquad c_p$ = specific heat of air of 0.24 Btu/lb/°F

$\qquad\qquad\qquad Q_c$ = flared gas heat release, Btu/hr

Chapter 11: Flare Stack Plume Rise 171

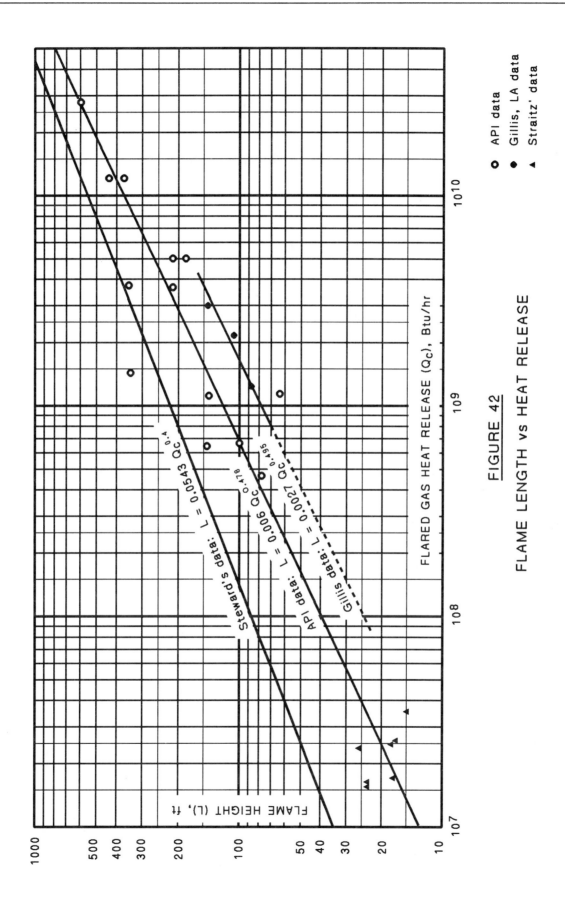

FIGURE 42

FLAME LENGTH vs HEAT RELEASE

Equation (69) can be simplified considerably by using an air density of 0.0749 lbs/ft³ at an air temperature of 70 °F, and by eliminating R from both sides of the equation:

(69a) $L = 0.8632 \, Q_c^{0.4} \, N'$

where: $N' = (r + w\rho_a/\rho)^{0.4} / [(NHV)^{0.4} (1 - w)]$

Evaluating values of N' for a wide range of flared gases (i.e., fuels), we obtain:

Fuel	r	NHV	ρ 100 °F	ρ 200 °F	N' 100 °F	N' 200 °F
H_2	34.5	52,020	0.0049	0.0042	0.059	0.059
CH_4	17.2	21,573	0.0392	0.0333	0.064	0.064
C_2H_6	16.1	20,476	0.0735	0.0624	0.063	0.063
C_2H_4	14.8	20,327	0.0686	0.0582	0.061	0.061
C_3H_8	15.7	19,985	0.108	0.0915	0.063	0.063
C_3H_6	14.8	19,733	0.103	0.0873	0.062	0.062
C_4H_{10}	15.4	19,719	0.142	0.121	0.063	0.063
C_4H_8	14.8	19,469	0.137	0.116	0.062	0.062
C_5H_{12}	15.3	19,551	0.176	0.150	0.063	0.063
C_5H_{10}	14.8	19,358	0.172	0.146	0.062	0.062
CO	2.5	4,348	0.0686	0.0582	0.055	0.055

The effect of the flared gas temperature upon the gas density (ρ) and, hence, upon the parameter N' is minimal within the range of 100 to 200 °F. Also, with the exception of carbon monoxide (CO), all of the N' values are within a range of 10 percent. For all practical purposes, N' can be taken to be 0.063 for the typical flared gas in a hydrocarbon processing plant. Therefore, equation (69a) can be further simplified to:

(70) $L = 0.0543 \, Q_c^{0.4}$

Steward's correlation in the simplified form of equation (70) has been included in Figure 42 for comparison with the API data and with Straitz's data.

Oenbring and Sifferman[67] have published data on flame lengths measured in a natural gas plant at Gillis, Louisiana. Their data is also included in Figure 42 and can be fit by this equation:

(71) $L = 0.0027 \, Q_c^{0.495}$

Oenbring and Sifferman suggest that their measured flame lengths at Gillis are predicted better by Brzustowski's method[68] than by the API data. Brzustowski's method is based upon the hypothesis that the end of the visible flame is the point where the flared gas fuel has been diluted to its lean flammability limit. However, comparing the Gillis data and the API data in Figure 42, the flame lengths measured at Gillis are only some 30-40 percent below the API data correlation line.

Upon considering all of the data in Figure 42, it appears that the API correlation as expressed by equation (68) provides an adequate basis for estimating flare stack flame lengths.

The API publication[64] also contains a method for estimating the flame tilt caused by the horizontal wind velocity impinging on the flame. The method is quite cumbersome and its theoretical rationale is not explained. This author suggests the simple and conservative approach of <u>assuming</u> that the flame is tilted 45° from the vertical. Thus, the vertical height vector of a flare stack flame becomes:

$$h_{fv} = L (\sin 45°) = 0.707 \, L$$

and using equation (68) for the flame length L, we obtain:

(72) $\qquad h_{fv} = 0.0042 \, Q_c^{0.478}$

Since the flame tilt has been assumed as 45° from the vertical, the horizontal downwind vector of the flame length (i.e., the flame reach) would be the same as the vertical height vector.

To give the reader some perspective, equation (72) provides these estimates of h_{fv}:

Q_c, Btu/hr	h_{fv}, ft
1×10^7	9
1×10^8	28
1×10^9	84
1×10^{10}	253

From the above estimates, it appears that the flame heights for flared gas heat releases of less than 1×10^8 Btu/hr are probably too small to be an important factor in determining the rise of combusted gas plumes from flare stacks. It also appears that, unless the flare heat release exceeds about 4×10^8 Btu/hr at which rate the flame height h_{fv} is only about 50 feet, there is probably very little need to consider the relative merits of obtaining the flame height by Brzustowski's method[68] vis-a-vis using equation (72) derived from the API data and the assumption of a 45° flame tilt.[†]

COMBUSTED GAS TEMPERATURES

The combusted gas temperature at the end of a flare stack flame is not required for determining the combusted gas plume rise, since the plume rise depends upon the Briggs buoyancy factor which can be obtained from the sensible heat emission of the plume (see Chapter 4).

However, if there is a need for obtaining the combusted gas temperature, it can be obtained by equating the combustion heat release from burning the flared gas (i.e., the fuel) with the

---

† This is not to say that Brzustowski's method may not be more accurate than the API data, but only that the much simpler API correlation is adequate for determining flare stack plume rises. For determining radiation heat fluxes from flares (for safety considerations), Brzustowski's method should certainly be considered since it does predict somewhat smaller flame lengths.

174   Chapter 11: Flare Stack Plume Rise

combusted gas heat content (enthalpy) at the combusted gas temperature. On a molar basis for a hydrocarbon fuel, that equality can be expressed as:

(73) $n_f (NHV) = \Sigma (n_1 H_1 + n_2 H_2 + n_3 H_3 + n_4 H_4)$

where: NHV = fuel net heating value at 298 °K (25 °C), cal/g-mol
n_f = g-mols of fuel
n_1 = g-mols of CO_2 in combusted gas
     = c
n_2 = g-mols of H_2O in combusted gas
     = 0.5h
n_3 = g-mols of N_2 in combusted gas
     = (1 + x)(3.76)(c + 0.25h)
n_4 = g-mols of oxygen in combusted gas
     = x(c + 0.25h)

c = carbon atoms per mol of fuel
h = hydrogen atoms per mol of fuel
x = (% excess air)/100

H_1, H_2, H_3, H_4 = enthalpies in cal/g-mol for the CO_2, H_2O, N_2 and O_2 in the combusted gas at the gas temperature T in °K

The enthalpies in equation (73) can be obtained from this general expression:

$$H = \int_{298}^{T} c_p \, dt$$

which translates to these expressions for the specific combustion gas components:

(74a) CO_2: $H_1 = 7.70\,T + (2.65 \times 10^{-3})\,T^2 - (0.28 \times 10^{-6})\,T^3 - 2522.5$

(74b) H_2O: $H_2 = 8.22\,T + (0.075 \times 10^{-3})\,T^2 + (0.447 \times 10^{-6})\,T^3 - 2468$

(74c) N_2: $H_3 = 6.76\,T + (0.305 \times 10^{-3})\,T^2 + (0.043 \times 10^{-6})\,T^3 - 2042.7$

(74d) O_2: $H_4 = 8.27\,T + (0.13 \times 10^{-3})\,T^2 + (1.88 \times 10^{5})\,T^{-1} - 3107$

where: H = cal/g-mol of heat content above 298 °K (25°C)
T = combusted gas temperature, °K

Calculation of combusted gas temperatures, with equations (73) and (74), involves trial-and-error selection of the temperature to find a value which satisfies the equality in equation (73).

If the combustion is stoichiometric (i.e., no excess air) and there is no heat transferred or lost from the combusted gas, the temperature obtained from equations (73) and (74) is referred to as the "adiabatic flame temperature at stoichiometric conditions". If heat is transferred or lost from the combustion gases, then the left-hand member of equation (73) must be reduced by the amount of heat transferred or lost.

The net heating values (NHV) of typical fuels found in flared hydrocarbon gases are:

Fuel Gas	NHV at 25 °C, cal/g-mol
Hydrogen	57,800
Methane	191,760
Ethane	341,260
Ethylene	316,200
Propane	488,530
Propylene	460,430
Butane	635,385
Butylenes	605,700
Pentane	782,040
Pentylenes	752,800

Table 22 lists calculated adiabatic flame temperatures for combusting various hydrocarbon fuel gases with no excess air and no thermal radiation losses. The calculations utilized the above fuel gas heating values and equations (73) and (74). As can be seen in Table 22, the adiabatic flame temperatures are well above 3500 °F. As discussed earlier in this chapter, that is the primary reason for defining flare stack plume rises as starting at the end of the visible flame ... where thermal radiation losses coupled with excess entrained air result in lowering the combusted gas temperature very considerably.

TABLE 22

CALCULATED ADIABATIC FLAME TEMPERATURES
(at zero percent excess air and with no radiation loss)

Fuel Gas Components	°F	°K
Hydrogen	4,014	2,485
Hydrogen:Methane, 1:1	3,801	2,367
Hydrogen:Methane, 1:2	3,776	2,353
Methane	3,740	2,333
Ethane	3,837	2,387
Ethylene	4,154	2,563
Propane	3,862	2,401
Propylene	4,055	2,508
Butane	3,873	2,407
Butylenes	4,012	2,484
Pentane	3,880	2,411
Pentylenes	3,992	2,473

Combusted gas temperatures were similarly calculated for various hydrocarbon fuel gases assuming that the flare stack flame entrains 100-175 percent excess air, and that the flame loses 25 percent of its heat release by thermal radiation. The results are plotted in Figure 43, and show that:

- At 100 percent excess air and 25 percent radiation losses, the combusted gas temperature from a hydrocarbon flare stack should be about 1745 to 1790 °F.

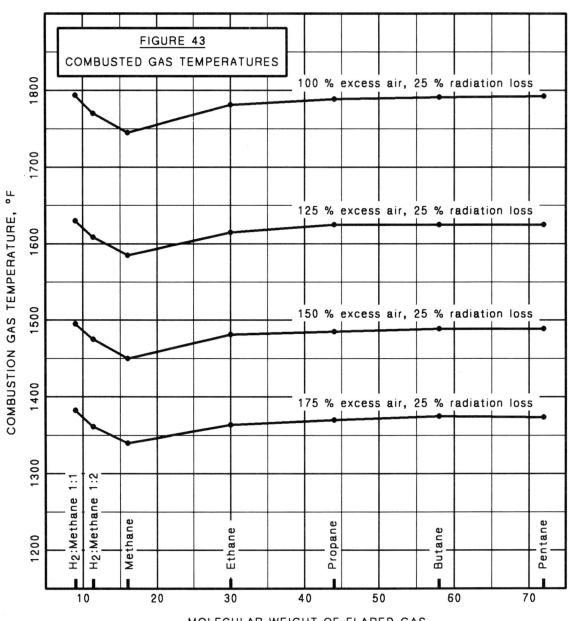

- At 175 percent excess air and 25 percent radiation losses, the combusted gas temperature from a hydrocarbon flare stack should be about 1340 to 1370 °F. For methane specifically, the temperature should be 1340 °F.

Straitz' flare test data[65] includes oxygen contents and temperatures **measured at the end of the visible flame**. His average results for burning methane in a 6-inch diameter flare were:

$$\begin{aligned}
\text{Oxygen content of combusted gas} &= 13 \text{ volume percent} \\
\text{(Equivalent percent of excess air)} &= 175 \text{ percent} \\
\text{Combusted gas temperature} &= 1295 \text{ °F}
\end{aligned}$$

Straitz' average measured temperature of 1295 °F at the visible flame tip is within 3 percent of the 1340 °F calculated for burning methane with 175 percent excess air and a 25 percent radiation loss (see Figure 43). **Thus, it appears that flare stack combusted gas temperatures at the end of the visible flame can be calculated by assuming 175 percent excess air and a 25 percent radiation loss.** It also appears that the visible flame tip temperature will be in the range of 1300 to 1400 °F for flares burning hydrocarbon gases ... which means that Briggs' plume rise equations can be applied to flare stack plumes with some degree of confidence.

CALCULATION OF FLARE STACK PLUME RISE

The key factor in determining the vertical height of a flare stack flame is the net heat released (Q_c) by the flared gas being burned at the stack exit.[†] The net heat released determines the flame length, from which the flame's vertical height (h_{fv}) is then obtained.

As developed earlier in this chapter, a flare stack flame may be assumed to dissipate about 25 percent of the net heat release as radiation heat loss.[††] Thus, the sensible heat emission available for imparting buoyancy to the combusted gas plume is about 75 percent of the net heat release. From that sensible heat emission, the Briggs buoyancy factor (F) can be determined for the combusted gas plume. The plume rise and the plume centerline at any downwind distance can then be calculated as discussed in Chapter 4.

Example 10 illustrates the calculation of the vertical height of a flare stack flame and the Briggs buoyancy factor for the combusted gas plume from the flame:

- Given the flared gas flow rate and composition, the net heat release Q_c is determined from the net heating values of the gas components.
- The vertical flame height h_{fv} is then calculated from the net heat release by using equation (72) developed in this chapter.
- The sensible heat emission Q, which imparts buoyancy to the combusted gas plume from the flare stack plume, is taken as 75 percent of the net heat release.
- The Briggs buoyancy factor is then calculated from the sensible heat emission Q by using equation (30b) developed in Chapter 4.
- The plume rise Δh and the plume centerline height H_e can then be determined.

[†] Based upon the net heating values rather than the gross heating values which include the heat expended in vaporizing water formed during combustion.

[††] Radiation losses may be 30-40 percent for burning gases of 44 molecular weight or more.

Chapter 11: Flare Stack Plume Rise

EXAMPLE 10: CALCULATION OF THE FLAME HEIGHT AND THE BUOYANCY FACTOR OF THE COMBUSTED GAS PLUME FROM A FLARE STACK

Calculate the vertical height (h_{fv}) of a flare stack flame, the Briggs buoyancy factor (F) of the combusted gas plume from the flame, and the plume centerline height (i.e., the stack effective height).

GIVEN: Flared gas flow rate = 1,000,000 SCF/hr

Flared gas composition:

Hydrogen	20 volume percent[†]
Methane	70 volume percent[†]
Ethane	8 volume percent[†]
Ethylene	2 volume percent[†]
	100

Flared gas molar flow rate
= (1,000,000 SCF/hr)(1 lb-mol/379 SCF)(454 g-mol/lb-mol)
= 1,200,000 g-mol/hr

Using net heating values tabulated earlier in this chapter:

Hydrogen = (0.20)(1,200,000 g-mol/hr)(57,800 cal/g-mol) = 13,872 × 10^6 cal/hr
Methane = (0.70)(1,200,000 g-mol/hr)(191,760 cal/g-mol) = 161,078 × 10^6 cal/hr
Ethane = (0.08)(1,200,000 g-mol/hr)(341,260 cal/g-mol) = 32,761 × 10^6 cal/hr
Ethylene = (0.02)(1,200,000 g-mol/hr)(316,200 cal/g-mol) = 7,589 × 10^6 cal/hr

Total net heat release, Q_c = 215,300 × 10^6 cal/hr
= 8.54 × 10^8 Btu/hr

From equation (72), the vertical flame height is:

h_{fv} = 0.0042(8.54 × 10^8)$^{0.478}$
= 78 feet

Assuming a 25 percent thermal radiation loss from the flame, the sensible heat emission is 75 percent of the net heat release:

Q = 0.75(8.54 × 10^8)
= 6.41 × 10^8 Btu/hr

From equation (30b), Briggs' buoyancy factor is:

F = (2.58 × 10^{-6})(6.41 × 10^8)
= 1,654 m^4/sec^3

With the buoyancy factor and other meteorological parameters, the combusted gas plume rise Δh can be calculated (see Examples 2, 3, and 4 in Chapter 4). The centerline height of the combusted gas plume is then:

$H_e = h_s + h_{fv} + \Delta h$
= stack height + flame height + plume rise

[†] Volume percent and mol percent are equivalent for a gas.

Chapter 12

MISCELLANY

SHORT-TERM RELEASES

Most plume dispersion models were developed for either a continuous gas flow from the source point or for an instantaneous puff release of very short duration ... but many short-term releases are neither continuous nor instantaneous. For example, emergency venting from pressure relief valves in an industrial plant may last from 1 or 2 minutes up to 10 or 15 minutes before corrective actions can be taken to stop the emergency venting.

The Federal Emergency Management Agency, the U.S. Department of Transportation and the U.S. Environmental Protection Agency have jointly distributed a handbook devoted to the analysis of chemical release hazards, which includes a software program called ARCHIE.[69] The ARCHIE program has a dispersion modelling module based upon the generalized Gaussian dispersion equation (13) for a continuous point-source plume incorporated with a "correction factor" for short-term continuous releases as developed by Palazzi et al.[70]

When the generalized Gaussian dispersion equation (13) for a continuous point-source plume has been corrected for short-term releases, the resulting plume dispersion equations for short-term continuous releases are:

$$(75) \quad C_{max} = \frac{C_c}{2} \left[\text{erf}\left(\frac{x}{1.414\,\sigma_x}\right) - \text{erf}\left(\frac{x - u t_r}{1.414\,\sigma_x}\right) \right] \quad \text{for: } x \leq \frac{u t_r}{2}$$

$$(76) \quad C_{max} = C_c \left[\text{erf}\left(\frac{u t_r}{2.828\,\sigma_x}\right) \right] \quad \text{for: } x > \frac{u t_r}{2}$$

The parameters in the above equations are:

C_{max} = Maximum concentration, g/m^3
C_c = Concentration obtained from the generalized Gaussian dispersion equation (13) for a continuous point-source plume, g/m^3
x = downwind distance to receptor, m
u = windspeed, m/sec
σ_x = downwind dispersion coefficient, m
  $\approx \sigma_y$ for short duration releases
t_r = duration of release, sec
erf = error function

The error function is defined as the integral of $(2/\sqrt{\pi})\,e^{-t^2}$ from $t = 0$ to $t = x$:

$$(77) \quad \text{erf}(x) = \frac{2}{\sqrt{\pi}} \int_0^x e^{-t^2} dt \quad \text{for: } x \geq 0$$

The erf (x) must be evaluated numerically and is approximated within 7 digits by:[71]

$$\text{erf}(x) = 1 - be^{-x^2} \quad \text{for: } x \geq 0$$

where: $b = 0.254829592\, a - 0.284496736\, a^2 + 1.421413741\, a^3 - 1.453152027\, a^4 + 1.061405429\, a^5$
$a = 1/(1 + 0.3275911\, x)$
$e = 2.71828$

For negative values of x, the following equality is used:

$$\text{erf}(-x) = -\text{erf}(x)$$

These are some values of erf (x) obtained with the above polynomial:

x	erf (x)
0.0	0
0.5	0.5205000
1.0	0.8427007
2.0	0.9953221
3.0	0.9999779
3.5	0.9999993
4.0	1.0000000
>4.0	1.0000000

All values of erf (x) lie within the range of 0 to 1, and all values of erf (-x) lie within the range of 0 to -1.

The generalized Gaussian dispersion equation (13) for a continuous point-source plume is repeated below for the sake of clarity:

$$(13) \quad C_c = \frac{Q}{u\, \sigma_z\, \sigma_y\, 2\pi}\, e^{-y^2/2\sigma_y^2} \left[e^{-(z_r - H_e)^2/2\sigma_z^2} + e^{-(z_r + H_e)^2/2\sigma_z^2} \right]$$

where: C_c = concentration of emissions, g/m³, at any receptor located at:
x meters downwind
y meters crosswind from the centerline
z_r meters above ground
Q = source emission rate, g/sec
u = windspeed, m/sec
H_e = plume centerline height above ground, m
σ_z = vertical dispersion coefficient, m
σ_y = crosswind dispersion coefficient, m

Thus, the expressions containing the error functions, by which C_c is multiplied in equations (75) and (76), are Palazzi et al's "correction factors" for short-term continuous releases, which take into account the dependence of the emissions concentration C on the duration of the release.

For those readers who are interested in the detailed mathematics, Palazzi et al derived equations (76) and (77) by starting with the well-known Gaussian dispersion equation for an instantaneous puff release:

(i) $$dC = \frac{dM}{\sigma_x \sigma_y \sigma_z (2\pi)^{1.5}} e^{-(x-ut)^2/2\sigma_x^2} e^{-y^2/2\sigma_y^2} \left[e^{-(z_r - H_e)^2/2\sigma_z^2} + e^{-(z_r + H_e)^2/2\sigma_z^2} \right]$$

where ut is the downwind distance travelled by the puff through the time t (starting at the instant of release).

A short-term continuous release of duration t_r at a constant flow of Q can be considered as being an infinite number of overlapping elementary puffs, where:

(ii) $\quad dM = Q\, dt_s \quad$ for: $\quad 0 \leq t_s \leq t_r$

where: M = instantaneous source emission strength, g
Q = short-term continuous source emission rate, g/sec
t_s = release starting time for any of the elementary puffs, sec

And the total concentration is the integral of equation (i):

(iii) $$C = \int_0^t \frac{Q}{\sigma_x \sigma_y \sigma_z (2\pi)^{1.5}} e^{-[x-u(t-t_s)]^2/2\sigma_x^2} e^{-y^2/2\sigma_y^2} \left[e^{-(z_r - H_e)^2/2\sigma_z^2} + e^{-(z_r + H_e)^2/2\sigma_z^2} \right] dt_s$$

Combining equation (iii) with equation (13):

(iv) $$C = C_c \int_0^t \frac{u}{\sigma_x (2\pi)^{0.5}} e^{-[x-u(t-t_s)]^2/2\sigma_x^2} dt_s \qquad \text{for: } t \leq t_r$$

(v) $$C = C_c \int_0^{t_r} \frac{u}{\sigma_x (2\pi)^{0.5}} e^{-[x-u(t-t_s)]^2/2\sigma_x^2} dt_s \qquad \text{for: } t \geq t_r$$

Let $\quad \mu = \dfrac{x - u(t - t_s)}{2^{0.5} \sigma_x} \quad$ to facilitate integration of equations (iv) and (iv):

$\quad\quad = \dfrac{x - ut}{2^{0.5} \sigma_x} \quad$ for: $\quad t_s = 0$

$\quad\quad = \dfrac{x}{2^{0.5} \sigma_x} \quad$ for: $\quad t_s = t$

$\quad\quad = \dfrac{x - u(t - t_r)}{2^{0.5} \sigma_x} \quad$ for: $\quad t_s = t_r$

Thus: $\quad \dfrac{d\mu}{dt_s} = \dfrac{u}{2^{0.5} \sigma_x} \quad$ and $\quad dt_s = \dfrac{2^{0.5} \sigma_x}{u} d\mu$

182 Chapter 12: Miscellany

Equation (iv) now becomes:

$$\text{(vi)} \quad C = C_c \int_{\frac{x-ut}{2^{0.5}\sigma_x}}^{\frac{x}{2^{0.5}\sigma_x}} \frac{u}{\sigma_x (2\pi)^{0.5}} \frac{\sigma_x 2^{0.5}}{u} e^{-\mu^2} d\mu$$

$$= \frac{C_c}{2} \frac{2}{\sqrt{\pi}} \int_{\frac{x-ut}{2^{0.5}\sigma_x}}^{\frac{x}{2^{0.5}\sigma_x}} e^{-\mu^2} d\mu$$

Utilizing the definition of the error function in equation (77):

$$\text{(vii)} \quad C = \frac{C_c}{2} \Big[\text{erf}(\mu) \Big]_{\frac{x-ut}{1.414\,\sigma_x}}^{\frac{x}{1.414\,\sigma_x}}$$

Which translates to:

$$\text{(viii)} \quad C = \frac{C_c}{2} \left\{ \text{erf}\left[\frac{x}{1.414\,\sigma_x}\right] - \text{erf}\left[\frac{x-ut}{1.414\,\sigma_x}\right] \right\} \qquad \text{for: } t \leq t_r$$

And, similarly, equation (v) becomes:

$$\text{(ix)} \quad C = \frac{C_c}{2} \left\{ \text{erf}\left[\frac{x-u(t-t_r)}{1.414\,\sigma_x}\right] - \text{erf}\left[\frac{x-ut}{1.414\,\sigma_x}\right] \right\} \qquad \text{for: } t \geq t_r$$

Equation (viii) reaches its maximum when $t = t_r$ (i.e., when the short-term release ends), and so equation (viii) becomes equation (75):

$$(75) \quad C_{max} = \frac{C_c}{2} \left[\text{erf}\left(\frac{x}{1.414\,\sigma_x}\right) - \text{erf}\left(\frac{x-ut_r}{1.414\,\sigma_x}\right) \right] \qquad \text{for: } x \leq \frac{ut_r}{2}$$

Equation (ix) reaches its maximum when its derivative vanishes, which is when:

$$\text{(x)} \quad e^{-\left[\frac{x-u(t-t_r)}{1.414\,\sigma_x}\right]^2} - e^{-\left[\frac{x-ut}{1.414\,\sigma_x}\right]^2} = 0$$

Thus, the maximum for equation (ix) occurs when:

$$\text{(xi)} \quad x - u(t-t_r) = \pm(x - ut)$$

Which corresponds to:

(xii) $t = \dfrac{t_r}{2} + \dfrac{x}{u}$ when equation (ix) reaches its maximum

Thus, equation (ix) becomes:

(xiii) $C_{max} = \dfrac{C_c}{2} \left\{ erf\left[\dfrac{u\,t_r}{2.828\,\sigma_x} \right] - erf\left[\dfrac{-u\,t_r}{2.828\,\sigma_x} \right] \right\}$

Since erf (-x) = - erf (x), equation (xiii) becomes equation (76):

(76) $C_{max} = C_c \left[erf\left(\dfrac{u\,t_r}{2.828\,\sigma_x} \right) \right]$ for: $x > \dfrac{u\,t_r}{2}$

Palazzi and his colleagues went beyond equations (75) and (76) to derive another set of equations for calculating the spatial distribution of concentrations from short-term releases averaged over different exposure times ... which are considerably more complex than are equations (75) and (76). If interested, the reader should study their publication.[70]

SOURCE AND RECEPTOR AT DIFFERENT GROUND-LEVEL ELEVATIONS

The Gaussian dispersion equation (13) includes a receptor elevation term z_r, which provides a mechanism for calculating plume concentrations at receptors located above ground-level. That mechanism is useful, for example, in determining plume concentrations on elevated structures or elevated working platforms.

When a ground-level receptor is on terrain at an elevation either lower or higher than the terrain at the base of the plume source stack, then the receptor ground-level concentrations may be approximated by adjusting the effective stack height H_e (the plume centerline height) to reflect the difference in terrain elevations between the source stack and the receptor:

(78)   Adjusted H_e = H_e ± terrain elevation difference
            = $h_s + \Delta h$ ± terrain elevation difference

Adjusting H_e is a simplistic "work-around" for the problem of varying terrain elevations. The reader would do well to review Egan's more sophisticated approach to this problem.[23]

MORE RIGOROUS DEFINITION OF BRIGGS' BUOYANCY PARAMETER

These equations were derived in Chapter 4 as definitions of the Briggs buoyancy parameter for hot stack gas plumes:

(29a) $F = (gV_s/\pi)(T_s - T_a)/T_s$

(29c) $F = gQ/(\pi c_{pa} T_a \rho_a)$

The derivation of equations (29a) and (29c) was based on the explicit assumption that the plume source stack gas had the same molecular weight (m) and the same specific heat (c_p) as that of air. In other words, $m_s = m_a$ and $c_{ps} = c_{pa}$, where the subscript s refers to the stack gas and the subscript a refers to air. That assumption is valid in most cases, since stack gases are usually the products of fuel combustion which do, in fact, have essentially the same molecular weight and specific heat as air.

However, it may be useful to consider those cases in which the stack gas does not have the same molecular weight and specific heat as air. In his 1975 lecture, Briggs presented a much more rigorous definition of his buoyancy parameter:[23]

(79) $\quad F = (g/\pi)(T_a/T_s)(V_s)(m_a - m_s)/(m_a) + (g/\pi)(Q)/(c_{pa}T_a\rho_a)$

The first term to the right of the equal sign in equation (79)[†] reflects the buoyancy difference between the surrounding ambient air and a stack gas due to the difference, if any, in their molecular weights (m_a and m_s). The second term reflects the buoyancy difference between the ambient air and stack gas due to the sensible heat content (Q) of the stack gas relative to the ambient air.

Equation (79) can be re-arranged so that both terms involve either the stack gas sensible heat emission Q or the stack gas volumetric flow rate V_s. The re-arrangement is made by using two equalities, the first one being:

(i) $\quad Q = \rho_s V_s c_{ps}(T_s - T_a)$

The second equality is obtained from the equation of state for ideal gases at constant pressure:

(ii) $\quad \rho_s/\rho_a = (T_a/T_s)(m_s/m_a)$

Using equations (i) and (ii), equation (79) can be re-arranged to either of these two forms:

(80) $\quad F = (g/\pi)(T_a/T_s)[(m_a - m_s)/m_a](Q)/[\rho_s c_{ps}(T_s - T_a)] + (g/\pi)(Q)/(c_{pa}T_a\rho_a)$

(81) $\quad F = (g/\pi)(T_a/T_s)(V_s)(m_a - m_s)/(m_a) + (g/\pi)(V_s)(m_s/m_a)(c_{ps}/c_{pa})(T_s - T_a)/(T_s)$

Both terms in the right-hand side of equation (80) involve Q ... whereas, both terms in the right-hand side of equation (81) involve V_s. **Equations (80) and (81) are both valid for the general case where $m_a \neq m_s$ and $c_{pa} \neq c_{ps}$.**

When $m_a = m_s$, equation (80) readily reduces to equation (29c). Likewise, when $m_a = m_s$ and $c_{pa} = c_{ps}$, equation (81) readily reduces to equation (29a). As noted earlier in this section, when the stack gases are the typical products of fuel combustion, their molecular weight and specific heat are essentially the same as that of the ambient air and equations (29a) and (29c) are valid. Otherwise, the generalized equations (80) or (81) should be used.

[†] The first term is zero when $m_a = m_s$ and that is presumably why Briggs did not include the first term in his earlier plume rise publication.[28]

CONVERTING PLUME CONCENTRATION EXPRESSIONS

The input parameters used in the Gaussian dispersion equations are usually expressed in dimensional units such that the calculated plume component concentrations are obtained as micrograms per cubic meter (i.e., $\mu g/m^3$). Since many of our ambient air pollutant limits are expressed in terms parts per million by volume (i.e., ppmv) by our regulatory agencies, one should be familiar with the numerical conversion of $\mu g/m^3$ to ppmv.

The EPA defines their reference conditions for gas volumes as being 25 °C at 1 atmosphere of pressure.[72] The generalized conversion between $\mu g/m^3$ and ppmv at those reference conditions is:

μg of gas component n per m^3 of total gas at 25 °C and 1 atmosphere pressure
= (40.883)(ppmv of gas component n)(molecular weight of n)

For SO_2 specifically, which has a molecular weight of 64.06:

1 ppmv of SO_2
= 2,619 μg of SO_2 per m^3 of total gas at 25 °C and 1 atmosphere pressure

For NO_2 specifically, which has a molecular weight of 46.01:

1 ppmv of NO_2
= 1,881 μg of NO_2 per m^3 of total gas at 25 °C and 1 atmosphere pressure

For CO specifically, which has a molecular weight of 28.01:

1 ppmv of CO
= 1,145 μg of CO per m^3 of total gas at 25 °C and 1 atmosphere pressure

It should be noted that the EPA's reference conditions for gas volumes (25 °C and 1 atmosphere) are <u>not</u> the same as in conventional engineering usage:

- In English units, the conventional reference conditions for a standard cubic foot (SCF) of gas are 60 °F and 1 atmosphere.

- In metric units, the conventional reference conditions used in most European nations for a normal cubic meter (Nm^3) of gas are 0 °C and 1 atmosphere.

Thus, 1 m^3 of gas at the EPA's 25 °C and 1 atmosphere is:

= 34.196 SCF at 60 °F and 1 atmosphere
= 0.916 Nm^3 at 0 °C and 1 atmosphere

EFFECT OF ALTITUDE ON AMBIENT AIR STANDARDS

Ambient air standards or pollutant limits, which are expressed as volumetric concentrations such as ppmv, are not dependent upon altitude. However, ambient air standards, expressed as mass per unit of gas volume at 1 atmosphere reference pressure, will vary with the altitude. For example, let us consider an air quality SO_2 standard of say 0.10 ppmv. In terms of $\mu g/m^3$ (mass per unit of gas volume) at sea level at 1 atmosphere of pressure, that standard translates to:

$$\mu g/m^3 = (0.10)(64.06)(44.803) \quad \text{(see previous section above)}$$
$$= 261.9 \text{ at sea level}$$

However, in an area at an altitude of 6,000 feet (such as northwest New Mexico), the atmospheric pressure is:

$$P = 14.696(0.963)^6 \quad \text{(see Chapter 1)}$$
$$= 11.72 \text{ psia}$$
$$= 0.7975 \text{ atmospheres}$$

and the SO_2 standard of 261.9 $\mu g/m^3$ at sea level becomes:

$$\mu g/m^3 = 261.9(0.7975)$$
$$= 208.9 \text{ at 6,000 feet altitude}$$

In other words, the standard of 0.10 ppmv is equivalent to 261.9 $\mu g/m^3$ at sea level and to 208.9 $\mu g/m^3$ at an altitude of 6,000 feet. Obviously, the effect of altitude on ambient air standards can be quite significant for many parts of the Western states.

CATEGORIZING ATMOSPHERIC STABILITY WITH THE RICHARDSON NUMBER

The stability parameter, s, is defined by equation (31) in Chapter 4 as:

$$s = (g/T_a) \, d\theta/dz$$

and it may be considered as being proportional to the rate at which convective heat transfer suppresses turbulence.

The parameter, $(du/dz)^2$, is the square of the rate at which the windspeed changes with altitude and it may be considered as being proportional to the rate at which mechanical shear in the atmosphere generates turbulence.

The Richardson number, Ri, is the dimensionless ratio of the stability parameter, s, to the parameter $(du/dz)^2$:

$$Ri = (g/T_a) \, (d\theta/dz)/(du/dz)^2$$

and, hence, it is a measure of the relative influence or importance of turbulence suppression by convective heat transfer as compared to turbulence generated by mechanical shear.[73] In any event, it has been used to categorize atmospheric stability as follows:[73]

Ri ≥ 0.25	Very stable
0 < Ri < 0.25	Stable
Ri = 0	Neutral
-0.03 < Ri < 0	Unstable
Ri ≤ -0.03	Very unstable

Chapter 12: Miscellany   187

MORE ON DISPERSION COEFFICIENTS IN EQUATION FORM

Chapter 2 contains a discussion of equation (27), developed by McMullen[17], for calculating Pasquill's rural dispersion coefficients. Table 7 in Chapter 2 demonstrates how very well equation (27) compares to Turner's graphs[9] of Pasquill's rural dispersion coefficients (see Figures 17 and 18 in Chapter 2).

Hanna and Drivas[41] present a set of equations (ascribed to Hanna, Briggs and Hosker[74]) to be used for the same purpose, namely the calculation of dispersion coefficients. Hanna and Drivas do not state whether the equations are intended to provide rural or urban dispersion coefficients, nor do they state that the equations are intended to represent the Pasquill dispersion coefficients. In any event, the Hanna, Briggs and Hosker equations[74] compare to McMullen's equation[17] as follows:

Pasquill Stability Class	Distance (km)	σ_z (meters)		
		Equation (27)	Hanna et al[74]	Figure 17
A	1.7	1383	340	1380
B	5.0	626	600	630
C	5.0	267	283	265
D	5.0	92	103	90
E	5.0	56	60	55
F	10.0	47	40	46

Pasquill Stability Class	Distance (km)	σ_y (meters)		
		Equation (27)	Hanna et al[74]	Figure 18
A	5.0	861	898	860
B	5.0	656	653	640
C	5.0	450	449	450
D	5.0	296	327	290
E	5.0	219	245	220
F	5.0	148	163	149

For a number of the above comparisons, the σ_z and σ_y values obtained from the equations of Hanna et al[74] differ considerably from the values obtained from McMullen's equation[17] and from Figures 17 and 18. In particular, at a distance of 1.7 km and at stability A conditions, the σ_z obtained from the equations of Hanna et al is one-fourth of the value obtained either from McMullen's equation or from Figure 17.

REFERENCES

(1) Pasquill, F., "Atmospheric Diffusion", 2nd Edition, John Wiley and Sons, New York, 1974

(2) Carpenter, S.B. et al, "Principal plume dispersion models, TVA power plants", JAPCA, August 1971

(3) Nuclear Regulatory Commission, "Safety guide 123, onsite meteorological programs", February 1972

Electric Power Research Institute, "Air quality models required data characterization", EPRI Report EC-137, May 1976

(4) Holzworth, G.C., "Mixing heights, wind speeds and potentials for urban air pollution throughout the contiguous United States", U.S. EPA Publication AP-101, 1972

(5) Environmental Protection Agency, "User's guide for the Climatological Dispersion Model", U.S. EPA Publication EPA-R4-73-024, December 1973

(6) Portelli, R.V., "Mixing heights, wind speeds and ventilation coefficients for Canada", Climatological Studies No. 31, Atmospheric Environment Service, Downsview, Ontario, Canada, 1977

(7) Slade, D.H. (editor), "Meteorology and atomic energy 1968", Air Resources Laboratories, U.S. Dept. of Commerce, July 1968

(8) Pasquill, F., "The estimation of the dispersion of windborne material", Meteor. Mag., February 1961

(9) Turner, D.B., "Workbook of atmospheric dispersion estimates", U.S. EPA Publication AP-26, revised 1970

(10) Hewson, E.W., Bierly, E.W., and Gill, G.C., "Topographic influences on the behavior of stack effluents", Proc. Amer. Power Conf., Vol. XXIII, 1961

(11) Bosanquet, C.H. and Pearson, J.L., "The spread of smoke and gases from chimneys", Trans. Faraday Soc., 32:1249, 1936

(12) Sutton, O.G., "The problem of diffusion in the lower atmosphere", QJRMS, 73:257, 1947

Sutton, O.G., "The theoretical distribution of airborne pollution from factory chimneys", QJRMS, 73:426, 1947

(13) Meade, P.J., "Meteorological aspects of the peaceful uses of atomic energy, Part I", Tech. Note 33, World Meteorological Organization, WMO No. 97.TP.41, 1960

(14) Gifford, F.A., "Use of routine meteorological observations for estimating atmospheric dispersion", Nuclear Safety, 2(4):47-51, 1961

(15) Bowne, N.E., "Diffusion rates", JAPCA, September 1974

(16) Environmental Protection Agency, "User's guide for HIWAY, a highway air pollution model", (specifically, see the computer program subroutine DBSTIG), U.S. EPA Publication EPA-650/4-74-008, February 1975

[The DBSTIG subroutine for generating dispersion coefficients is utilized in many of the EPA's UNAMAP dispersion models.]

(17) McMullen, R.W., "The change of concentration standard deviations with distance", JAPCA, October 1975

(18) McElroy, J.L. and Pooler, F., "The St. Louis dispersion study, Volume II-Analysis", U.S. EPA Publication AP-53, December 1968

(19) McElroy, J.L., "A comparative study of urban and rural dispersion", J. Appl. Meteor., 8(1):19, 1969

(20) Shum, Y.S., Loveland, W.D. and Hewson, E.W., "The use of artificial activable trace elements to monitor pollutant source strengths and dispersal patterns", JAPCA, November 1975

(21) Trinity Consultants, "Atmospheric diffusion notes", February 1977

(22) Briggs, G.A., "Diffusion estimates for small emissions", Air Resources Atmospheric Turbulence and Diffusion Laboratory, NOAA, ADTL-106, Oak Ridge, TN, 1973

(23) Haugen, D.A. (editor), "Lectures on air pollution and environmental impact analyses", Amer. Meteor. Soc., September 1975

(24) Bosanquet, C.H., Carey, W.F. and Halton, E.M., "Dust deposition from chimney stacks", Proc. Instit. Mech. Engrs., 162(3), 1950

(25) Bosanquet, C.H., "The rise of a hot gas plume", J. Instit. of Fuel, June 1957

(26) Briggs, G.A., "A plume rise model compared with observations", JAPCA, 15:433-438, 1965

(27) Briggs, G.A., "CONCAWE meeting: discussion of the comparative consequences of different plume rise formulas", Atmos. Envir., 2:228-232, 1968

(28) Briggs, G.A., "Plume rise", USAEC Critical Review Series, 1969

(29) Briggs, G.A., "Some recent analyses of plume rise observation", Proc. Second Internat'l. Clean Air Congress, Academic Press, New York, 1971

(30) Briggs, G.A., "Discussion: chimney plumes in neutral and stable surroundings", Atmos. Envir., 6:507-510, 1972

(31) Brummage, K.G. et al, "The calculation of dispersion from a stack", Stichting CONCAWE, The Hague, The Netherlands, 1966

(32) Holland, J.Z., "A meteorological survey of the Oak Ridge area", USAEC Report ORO-99, 1953

(33) Smith, M.E., "Recommended guide for the prediction of the dispersion of airborne effluents", ASME, 1968

(34) Moses, H. and Kraimer, M.R., "Plume rise determination - a new technique without equations", JAPCA, August 1972

(35) Slowik, A.H., Austin, J.M. and Pica, G.N., "Plume dispersion modeling in complex terrain under stable atmospheric conditions", Paper 77-29.1, 70th Annual APCA Meeting, Toronto, 1977

(36) England, W.G., Teuscher, L.H. and Snyder, R.B., "A measurement program to determine plume configurations at the Beaver Gas Turbine Facility", JAPCA, October 1976

(37) Turner, D.B., "Dispersion estimate suggestion No. 2 (revised)", Model Application Branch, U.S. EPA, May 1973

(38) Briggs, G.A., "Plume rise buoyancy effects", Atmospheric Science and Power Production, U.S. Dept. of Energy, Report TIC-27601, 1984

(39) Montgomery, T.C. and Coleman, J.H., "Empirical relationship between time-averaged SO_2 concentrations", Envir. Sci. & Tech., October 1975

(40) Trinity Consultants, "Atmospheric diffusion notes", Fall 1976

(41) Hannah, S.R. and Drivas, P.J., "Guidelines for the use of vapor cloud dispersion models", Center for Process Safety, Amer. Instit. of Chem. Engrs., 1987

(42) Nonhebel, G., "Recommendations on heights for new industrial chimneys", J. Instit. of Fuel, October 1960

(43) Environmental Protection Agency, "Air pollution meteorology", Air Pollution Training Institute Course 411, September 1975

(44) DeMarrais, G.A., "Wind speed profiles at Brookhaven National Laboratory", J. Appl. Meteor., 16:181, 1959

(45) Davenport, A.G., "The relationship of wind structure to wind loadings", Internat'l. Conf. on the Wind Effects on Buildings and Structures, National Physical Laboratory, England, June 1963

(46) Environmental Protection Agency, "User's guide for PAL, a Gaussian-plume algorithm for Point, Area and Line sources", U.S. EPA Publication EPA-600/4-78-013, February 1978

References

(47) Touma, J.S., "Dependence of the wind profile power law on stability for various locations", JAPCA, September 1977

(48) "Atmospheric dispersion modeling, a critical review", JAPCA, September 1979

(49) Bowne, N.E. et al, "Overview, results, and conclusions for the EPRI plume model validation and development project: Plains site", Electric Power Research Institute Final Report 1616-1 for Project EA-3704, 1983

(50) Benarie, M.M., "Editorial: The limits of air pollution modelling", Atmos. Envir., 21:1-5, 1987

(51) Ellis, H.M. et al, "Comparison of predicted and measured concentrations for 58 alternative models of plume transport in complex terrain", JAPCA, June 1980

(52) American Petroleum Institute, "An evaluation of short-term air quality models using tracer study data", API Report No. 4333, October 1980

(53) Beychok, M.R., "How accurate are dispersion estimates?", Hydrocarbon Processing, October 1979

(54) Bierly, E.W. and Hewson, E.W., "Some restrictive meteorological conditions to be considered in the design of stacks", J. Appl. Meteor., September 1962

(55) Novak, J.H. and Turner, D.B., "An efficient Gaussian-plume multiple-source air quality algorithm", JAPCA, June 1976

(56) Yamartino, R.J., "A new method of computing pollutant concentrations in the presence of limited vertical mixing", JAPCA, May 1977

(57) American Petroleum Institute, "Gaussian dispersion models applicable to refinery emissions", API Publication 952, October 1977

(58) Pooler, F., "Potential dispersion of plumes from large power plants", U.S. Public Health Service Publication 999-AP-16, 1965

(59) Pasquill, F., "Atmospheric dispersion parameters in Gaussian plume modeling, Part II: Possible requirements for change in Turner workbook values", U.S. EPA Publication 600/4-76-030b, June 1976

(60) Weast, R.C. (editor), "Handbook of Chemistry and Physics", 56th Edition, CRC Press, 1975-1976

(61) Gutfreund, P.D., "Limited mixing and inversion breakup evaluations using routine upper air and surface data", 68th Annual APCA Meeting, Boston, June 1975

(62) Turner, D.B., "A diffusion model for an urban area", J. Appl. Meteor., 3(1):83-91, 1964

(63) National Climatic Data Center, "STAR (Stability Array) - Normalized data", Documentation Manual TD-9773, Asheville, NC

(64) American Petroleum Institute, "Guide for pressure relief and depressuring systems", API RP 521, First Edition, September 1969

(65) Straitz, J.F., "Flaring for gaseous control in the petroleum industry", 71st Annual APCA Meeting, Houston, June 1978

(66) Steward, F.R., "Prediction of the height of turbulent diffusion buoyant flames", Combust. Sci. and Tech., Vol. 2 (203-212), 1970

(67) Oenbring, P.R. and Sifferman, T.R., "Flare design based on full-scale plant data", API 45th Midyear Refining Meeting, Houston, May 1980

(68) Brzustowski, T.A., "A model for predicting the shapes and lengths of turbulent diffusion flames over elevated industrial flares", 22nd Canadian Chemical Engineering Conference, Toronto, 1972

(69) "Handbook of chemical hazard analyses procedures", Federal Emergency Management Agency, U.S. Dept. of Transportation, and U.S. Environmental Protection Agency, 1989

"Automated Resource for Chemical Hazard Incident Evaluation (ARCHIE)", a computer program distributed with the above handbook.

(70) Palazzi, E. et al, "Diffusion from a steady source of short duration", Atmos. Envir., 16:2785-2790, 1982

(71) Abramowitz, M. and Stegun, I.A. (editors), "Handbook of Mathematical Functions With Formulas, Graphs and Mathematical Tables", National Bureau of Standards, Applied Mathematics Series 55, November 1964

(72) Code of Federal Regulations, Title 40, Part 50, Section 50.3

(73) Schnelle, K.B., "Atmospheric diffusion modeling", Encyclopedia of Physical Science & Technology, Volume 2, Academic Press Inc., 1987

(74) Hanna, S.R., Briggs, G.A. and Hosker, R.P., "Handbook on atmospheric diffusion", U.S. Department of Energy, DOE/TIC-11223, 1982